ARID LAND HYDROGEOLOGY: IN SEARCH OF A SOLUTION TO A THREATENED RESOURCE

DEVELOPMENTS IN ARID REGIONS RESEARCH SERIES

The aim of DARE's series is to present basic and applied research related to geoengineering problems in arid lands. Areas of interest include soil science and engineering in arid lands, information technology application to geoengineering, geoenvironmental engineering, hydrogeology, applied geomechanics, novel geotechnical construction techniques, sustainable land use, and case histories.

In the series, the books are intended for use in university courses at postgraduate level and are expected to be valuable reference materials for practicing engineers.

A.M.O. Mohamed
Series Editor

BOOKS IN SERIES

1. A.M.O. Mohamed & K.I. AL Hosani, K.I. (eds.) "*Geoengineering in Arid Lands*" 2000.
   A. A. Balkema Publishers, ISBN 90 5809 160 0, 720 pages.
2. A.M.O. Mohamed, D. Chenaf & S. El-Shahed "*DARE's Dictionary of Environmental Sciences and Engineering: English-French-Arabic*" 2003.
   A. Balkema Publishers, ISBN 90 5809 617 3, 556 pages.
3. A.M.O. Mohamed (ed.) "*Reclaiming the Desert: Towards a Sustainable Environment in Arid Lands*" 2006.
   Taylor & Francis/Balkema, ISBN 0 415 41128 9, 318 pages.
4. A.M.O. Mohamed (ed.) "*Arid Land Hydrogeology: In Search of a Solution to a Threatened Resource*" 2006.
   Taylor & Francis/Balkema, ISBN 0 415 41127 0, 198 pages.

Developments in Arid Regions Research, 4

PROCEEDINGS OF THE 3RD UAE – JAPAN SYMPOSIUM ON SUSTAINABLE GCC ENVIRONMENT AND WATER RESOURCES (EWR2006), ABU DHABI UNITED ARAB EMIRATES, 28–30 JANUARY, 2006

Arid Land Hydrogeology: In Search of a Solution to a Threatened Resource

Edited by

A.M.O. Mohamed
UAE University, Al Ain, United Arab Emirates

Taylor & Francis
Taylor & Francis Group

LONDON / LEIDEN / NEW YORK / PHILADELPHIA / SINGAPORE

Published by: Taylor & Francis/Balkema
P.O. Box 447, 2300 AK Leiden, The Netherlands
e-mail: Pub.NL@tandf.co.uk
www.balkema.nl, www.tandf.co.uk, www.crcpress.com

ISBN10 0-415-41127-0 ISBN13 978-0-415-41127-1

Printed in Great Britain

Arid Land Hydrogeology: In Search of a Solution to a Threatened Resource – Mohamed(ed.)
© 2006 Taylor & Francis Group, London, ISBN 0 415 41127 0

Table of Contents

EWR2006

Organized by the Research Affairs, UAE University and Japan Cooperation Center, Petroleum (JCCP), Japan.

UAE University

The *United Arab Emirates University* was established in 1976 by H. H. Sheikh Zayed Bin Sultan Al Nahayan, the father of the nation, as the flagship institution of higher education in the region and to realize his vision of knowledge based society. After almost three decades, the UAE University is standing tall as the premier national university whose mission is to meet the educational and cultural needs of the UAE society by providing programs and services of the highest quality. It contributes to the expansion of knowledge by conducting quality research and by developing and applying modern information technology and is recognized as the major research engine of the region. It plays a significant role in leading cultural, social and economic development in the country. UAEU maintains excellent teaching, an effective learning environment with an optimum student population size, quality applied research, and outstanding community services. The focus of the UAEU has also been on international recognition of excellence through accreditation and external evaluation of all its activities.

Japan Cooperation Center, Petroleum (JCCP)

Japan Cooperation Center, Petroleum (JCCP) was founded in 1981 under the auspices of the Ministry of Economy, Trade and Industry and with support from Japanese Oil and Engineering Industries. Its purpose is to work for mutual understanding between Japan and Oil-producing countries, to build relations of trust and friendship with those countries and to strengthen cooperative relationship among them. Oil-producing countries have needs in refining, marketing, and technological and human resources development. JCCP activities include: programs to accept downstream sector of the oil industry, which JCCP helps to satisfy through cooperation in overseas training and technical cooperation, etc.

The Third Joint
UAE - JAPAN Symposium
EWR
2006
Sustainable GCC Environment
and Water Resources
Jan. 28-30th 2006
Abu Dhabi, UAE
UAE University,
United Arab Emirates
Japan Cooperation
Center, Petroleum, Japan

Arid Land Hydrogeology: In Search of a Solution to a Threatened Resource – Mohamed(ed.)
© 2006 Taylor & Francis Group, London, ISBN 0 415 41127 0

About the Editor

Advisor to the Assistant Provost for Research, UAE University
Professor of Geotechnical and Geoenvironmental Engineering
Department of Civil and Environmental Engineering, UAE University
P.O. Box 17551, Al Ain, United Arab Emirates
Tel.: (9713) 7675580; Fax: (9713) 7675582;
E-mail: Mohamed.A@uaeu.ac.ae
Web: http://www.engg.uaeu.ac.ae/mohamed.a

A.M.O. Mohamed earned his M. Eng. in 1983 and Ph.D. in 1987 from the Department of Civil Engineering and Applied Mechanics, McGill University, Montreal, Canada. From 1987 to 1998 Dr. Mohamed was employed by McGill University, and was the *Associate Director* of the former Geotechnical Research Centre (GRC) and *Adjunct Professor* in the Department of Civil Engineering and Applied Mechanics. In 1998, he joined the Department of Civil and Environmental Engineering, UAE University, where he is currently *Professor of Geotechnical and Geoenvironmental Engineering*, and the *Advisor to the Assistant Provost for Research*.

Professor Mohamed is currently the *President* of the Gulf Society for Geoengineering (GSGE): http://www.engg.uaeu.ac.ae/gsge, the *Editor-in-Chief* of Developments in Arid Regions Research (DARE) Series Published by A.A. Balkema Publishers: http://balkema.ima.nl/Scripts/ cgiBalkema.exe, *Regional Editor*, Journal of Applied Sciences, Pakistan; *Executive Board Member*, Arab Healthy Water Association; http://www.mgwater.com/executive.shtml; *A member of the Geotechnical and Geoenvironmental Engineering Technical Committee* of International Society of Offshore and Polar Engineering (ISOPE); and the former *Editor-in-Chief* of the Emirates Journal for Engineering Research Published by UAE University. Prof. Mohamed also served as a *member of the International Advisory Board* for various international conferences.

Professor Mohamed's research activities have resulted in co-authoring the following three books:

1. Yong, R.N., Mohamed, A.M.O., and Warkentin, B.P. (1992) "*Principles of Contaminant Transport in Soils*," Elsevier, The Netherlands, 327 pages; the book was *translated into Japanese in 1995*.
2. Mohamed, A.M.O., and Antia, H.E. (1998) "*Geoenvironmental Engineering*," Elsevier, The Netherlands, 707 pages and
3. Mohamed, A.M.O., Chenaf, D., and El-Shahed, S. (2003) "*DARE's Dictionary of Environmental Sciences and Engineering: English-French-Arabic*, A.A. Balkema Publishers, ISBN 90 5809 617 3, 556 pages. http://balkema.ima.nl/Scripts/cgiBalkema.exe.

Professor Mohamed has edited/co-edited the following four books:

1. Yong, R.N., Hadjinicolaou, J., and Mohamed, A.M.O. (1993) "*Environmental Engineering*, Volumes 1 and 2, ASCE-CSCE
2. Mohamed, A.M.O., and Al-Hosani, K. (2000) "*Geoengineering in Arid Lands*," A. A. Balkema, ISBN 90 5809 160 0, 720 pages.
3. A.M.O. Mohamed (ed.) (2006) "*Arid Land Hydrogeology: In Search of a Solution to a Threatened Resource*" Taylor & Francis/Balkema and

4. A.M.O. Mohamed (ed.) (2006) "*Reclaiming the Desert: Towards a Sustainable Environment in Arid Lands*," Taylor & Francis/Balkema.

For publications in refereed journals and conference proceedings, Professor Mohamed published 190 ***papers***, and for training of highly qualified personnel, he *supervised 28 graduate students*(M. Sc., M. Eng. and Ph.D.) in UAE and McGill Universities.

Professor Mohamed's present research activities contribute to soil properties and behaviour, ground improvement, soil-pollutant interactions, transport processes, multi-phase flow, remediation of polluted soils, monitoring of subsurface pollutants via time domain electrometry techniques, waste management, environmental impact assessment and risk management.
Professor Mohamed awarded:

1. *The Outstanding Performance and Distinction Award in Research* in the College of Engineering, UAE University (2004),
2. *The* Outstanding *Performance and Distinction Award in University and Community Services* in the College of Engineering, UAE University (2004),
3. *The Best Interdisciplinary Research Project Award* in the College of Engineering, Research Affairs Sector, UAE University (2004),
4. *The Decree of Merit for Outstanding Contribution to the Geoengineering Field* from International Biographical Center (IBC), Cambridge, England (2002), and
5. *The Outstanding Performance and Distinction Award in Research* in the College of Engineering, UAE University (2001).

For Professor Mohamed's noteworthy achievements in his field, his biography has been included in Marquis Who's Who in Science and Engineering, 2002-present, Marquis Who's Who in the World, 20th Edition, 2002, and the International Biographical Center (IBC) inaugural edition of One Thousand Great Scientists, Cambridge, England (2003).

Arid Land Hydrogeology: In Search of a Solution to a Threatened Resource – Mohamed(ed.)
© 2006 Taylor & Francis Group, London, ISBN 0 415 41127 0

Preface

For the last 13 years, Japan Cooperation Center, Petroleum (JCCP), (1992–2001) as PEC, has been organizing annual symposia to promote the enhancement of the environment under desert conditions, focusing on greenery development. These symposia have been held in the Gulf Cooperation Council (GCC) countries such as Kingdom of Saudi Arabia, State of Kuwait, United Arab Emirates (UAE), Sultanate of Oman and the Kingdom of Bahrain, as well as in Japan, and have attracted considerable interest from the scientists and professionals in those countries. In bringing together these scientists, the symposia provided forums for exchanging views and information on multitudes of issues ranging from water purification, usage and water management, to desert greening.

Following the success of the first and second UAE-Japan symposium at the UAE, this third joint UAE-Japan Symposium on sustainable GCC Environment and Water Resources (January 28–30th) held in Abu Dhabi, is jointly sponsored by the Research Affairs, UAE University, and Japan Cooperation Center, Petroleum (JCCP), Japan.

The set objectives of the symposium were:

1. To bring together scientists and professionals from GCC and Japan to communicate and exchange knowledge on arid land hydrogeology: in search of solution to a threatened resource, and
2. To provide better understanding of the land environment and rehabilitation of a threatened resource through utilization and application of innovative technologies.

Included in this book is a very good and comprehensive collection of contributions from prominent researchers and authorities, dealing with the most pressing problems in arid land hydrogeology: in search of solution to a threatened resource. Research studies and actual experiences dealing with ground water management, assessment, protection and sustainable water resources are some of the many research and application topics that have been addressed. Soil enhancement techniques for moisture control in arid lands are also discussed in this volume.

The success of the symposium and the quality of the papers presented in this volume have been due not only to the tireless effort of the members of the organizing committee, but also to the many contributions made by the members of the Research Affairs sector, UAE university.

Although many excellent abstracts were unfortunately judged "unsuitable" for the symposium and were not included in the program, acknowledgement and appreciation should be extended to those authors for their interest in this symposium. To many paper presenters and session chairpersons, we owe our considerable gratitude for their efforts and participation. This volume proceeding of the third joint UAE-Japan symposium bears testimony to their magnificent efforts and contributions and will serve to preserve the record of arid land hydrogeology: in search of solution to a threatened resource as represented by UAEU and JCCP.

A.M.O. Mohamed
2006

Arid Land Hydrogeology: In Search of a Solution to a Threatened Resource – Mohamed(ed.)
© 2006 Taylor & Francis Group, London, ISBN 0 415 41127 0

Summary, recommendations and suggestions for actions

A.M.O. Mohamed
Department of Civil and Environmental Engineering, UAE University, UAE

1 SUMMARY AND RECOMMENDATIONS

Papers presented at the Third Joint UAE-JAPAN Symposium on Sustainable GCC Environment and Water Resources, EWR2006, January 28–29, 2006, Abu Dhabi, UAE, was divided into two themes.

Theme A related to the environmental issues and published in "*Reclaiming the Desert: Towards a Sustainable Environment in Arid Lands*" In Developments in Arid Regions Research, DARE, Vol. 3, A.M.O. Mohamed (ed.), Taylor & Francis/Balkema, 2006.
Theme B related to arid land hydrogeological issues and published in "*Arid Land Hydrogeology: In Search of a Solution to a Threatened Resource*" In Developments in Arid Regions Research, DARE, Vol. 4, A.M.O. Mohamed (ed.), Taylor & Francis/Balkema, 2006.

After intensive discussions conducted by the participants of EWR 2006, the most important areas that have been selected for recommendations are:

1. Soil management;
   a. Sulfur utilization in agriculture
   b. Solar distillation
   c. Soil moisture enhancement
   d. Use of oily treated water in agriculture
   e. Soil solarization for direct thermal inactivation of soil borne-pests
   f. Utilization of adopted indigenous forage species to combat drought during vegetative and reproductive stages in the field
2. Water management challenges
3. Groundwater recharge in arid lands;
4. Groundwater contamination and management;
5. Monitoring systems;
6. Bioreactor Landfills;
7. Treatment of contaminated soils;
8. Waste water treatment
9. Environmental impact assessment;
10. Risk management of polluted sites; and
11. Sulfur utilization in construction industry
12. Sustainable cities
13. Global warming
14. Oil spill management

1.1 *Soil management*

The constraints on managing soils for sustainable land use in dry lands are different in nature. Some are connected with the specificity of the natural conditions including soils, others with the prevailing socioeconomic conditions. Appropriate management of soils is the prerequisite of sustainable land use in the dry lands. Certain problems still require extensive research, e.g., soil resilience, particularly in view of the diversity of dry land soils in different ecological conditions. Another area of research in the dry lands is the relationship between the processes of soil degradation and the resultant different ecological situations; to answer the question whether soil erosion is always harmful or whether it may be beneficial in certain cases depending on the nature of the soil parent material and soil formation. The above problems concerning the soils of dry lands are recommended as priorities for fundamental and adaptive research by the appropriate scientific institutions.

Alkaline soil widely spreads in arid regions. This alkaline nature of soil is a reason for the scarce vegetation in these regions, although a lack of water is also a factor. Soil pH plays a major part in the volatilization of microelements in soil, and this influences the growth and development of vegetation. Although the optimum pH range is generally considered to be 5.5–6.5, it is not rare that soil pH exceeds 8.0 in the Middle East region.

A lack of organic matters causes another problem. Organic matters in the soil play an important role in the agglomeration fertility of the soil, and the slow and continuous release of nutrition to plants. Under normal circumstances with healthy vegetation, microorganisms and animals as well as plants that inhabit the same environment contribute to a sound organic cycle. Without this cycle in place, however, mineralization generally progresses rapidly in arid regions.

a. *Sulfur utilization in agriculture*

The utilization in sulfur is a well-known method for reducing the pH level in alkaline soil. In interaction with sulfur oxidizing bacteria, sulfur is oxidized into sulfuric acid, which neutralizes alkaline soil. Generally, sulfur oxidizing bacteria are found in natural soil. However, because micro flora is scarce in the desert highly mineralized soils new technologies are needed for the utilization of the high performance species isolated and cultured in advance and then re-introduced into the soil together with the sulfur.

Moreover, sulfur is an essential nutrient for plants and is generally absorbed as sulfate by plants. Therefore, technologies to transform elemental sulfur into sulfate ions would not only contribute to the improvement of soil pH but also to the soil benefits by supplying essential sulfur nutrients.

In order to resolve the issue of a lack of organic matter, varying composts or peat moss are generally used. These methods, however, are not cost effective. For this reason, new technologies such as the use of excess sludge generated from the sewage treatment process as the organic matter should be looked at and utilized. The efficacy of sludge as an organic fertilizer has long been recognized in the world, and its use has proven successful in nurturing fertile farmland. In addition, it is beneficial to consider the reuse of the excess sludge since sewage treatment technologies are common throughout the world and the sludge can easily be obtained.

Reported studies have focused on the possible application of sulfur in plant cultivation through a combination of the other technologies such as sludge generated from waste treatment plants. Emphasis were given to the use of materials and techniques that can be procured locally, and combination of practical technologies to create a single soil amendment in view of convenient applications.

b. *Solar distillation*

Due to environmental, ecological and economical activities, particularly in an arid region (such as Middle East) and remote areas, fresh water demand mainly depends on desalination performance produced from brackish or sea water. However, such region is rich in solar energy with high intensity so that solar desalination may be an ideal solution. Solar distillation is the simplest desalination technique, compared to others such as multiple-effect distillation, multi-stage flash, reverse osmosis, electro-dialysis and biological treatment. A basin-type solar still is the most popular method of solar distillation but has undergone very little advances due to low distillate productivity and the difficulty of rapid and easy removal of salt accumulation in the basin. More studies are needed to overcome design problems and efficiency. Other techniques such as "Tubular Solar Still" should be looked at.

c. *Soil moisture enhancement*

Agricultural polymers are known since the 1950's and their uses were limited because of the high prices. But, recent retail prices became more affordable due to extensive research and generated private sectors on this new technology. For instance, highly crossed-polyacrylamide potassium based hydrophilic gel, was found useful to improve water absorption capability and availability of sand dunes for medium saline water.

Most of cultivated farms in Gulf Cooperation Council (GCC) countries depend on use of ground water as source for irrigation, where ground water is a limited resource and due to continuous pumping by huge quantities for irrigation and almost zero recharging, dramatic annual depletion is unavoidable. Introduction of new techniques of water conservation in the agriculture industry is a must water and energy conservation. One of the recent promising technologies is use of artificial soil conditioners, such as polymers. More research is required to evaluate the required dosage, optimum environmental conditions and environmental impact.

d. *Use of oily treated water in agriculture*

During oil production, large quantities of water are also produced. For example, Petroleum Development Oman (PDO) in 2002 produced some 330,000 m^3/day of water as a by-product of its oil output of approximately 135,000 m^3/day. The volume of this production water has been steadily increasing over time and was predicted to rise to about 650,000 m^3/day by the year 2005, as oil fields get older. Approximately 37% (123,000 m^3/day) of the currently produced water is utilized for reservoir pressure maintenance by injection mainly into northern Oman fields. About 140,000 m^3/day production water is presently disposed off into shallow (150–400 m) sub-surface formation at oil field sites mainly in south Oman.

According to the new environmental regulations in GCC countries, this practice should be stopped. The disposal of the produced water to deep formations is expensive. One of the possible alternatives to the underground disposal of produced water is land disposal by using it as irrigation water for salt tolerant crops. Special treatments are needed to clean up the produced water from its oil content. More studies are needed to assess the effect of treated oil production water on soil properties due to land disposal as well as on crop production and toxicity. Above all, risk assessment to human health and the environment is a must.

e. *Soil solarization for direct thermal inactivation of soil borne-pests*

The substantial expansion in agricultural production in United Arab Emirates (UAE) in recent years has been associated with development and spread of several weeds and other soil-borne pathogens. Methyl bromide, the sodium salt of metham and dazomet is widely used in the United Arab Emirates for soil fumigation in open fields and in green houses. However, these fumigants are considered to be harmful for all organisms. The loss of methyl bromide fumigant to the air and possibly reaching groundwater as a contaminant and its potential ozone reduction has stimulated interest in finding an alternative method for disinfestation of soil–borne pests and weeds. Soil solarization was considered as appropriate, non chemical, alternative to methyl bromide fumigation.

Soil solarization is a natural, hydrothermal process of disinfesting soil of plant pests by using passive solar heating of moist soil mulched with polyethylene sheets. This simple technique has the potentiality to increase the crop yield and improve soil chemical characters. In addition, this environmental friendly technique does not have negative effects on many of beneficial microorganisms such as arbuscular mycorrhizal fungi and nitrogen fixing bacteria.

The effectiveness of soil solarization for direct thermal inactivation of soil borne-pests is dependant on a number of physical factors including solar radiation intensity and duration, air temperature, amount of humidity at the soil surface beneath the tarps, properties of plastics used to produce the green house effect resulting in the heating of soil and properties of the soil to be treated. More research is needed to evaluate the role of these controlling parameters on the success of the technology.

f. *Utilization of adopted indigenous forage species to combat drought during vegetative and reproductive stages in the field*

The Arabian Peninsula experiences some of the most extreme climatic conditions found on the Earth. It is characterized by low erratic rainfall, high evaporation rates, extremely high temperatures during summer, and high soil salinities. A sustainable livestock industry in the Arabian Peninsula requires sustainable systems of both grazing and the production of cheap fodder plants. Currently, the main fodder crops in the area are alfalfa and Rhodes grass, which are not adapted the prevailing conditions and require vast quantities of water (up to 48,000 m^3/ha/yr), which often derived from non-renewable groundwater sources. The production of these forages has resulted in abandonment of many farms due to the increase in soil salinity. The utilization of adopted indigenous forage species to replace the exotic ones could be the solution for such problem. *Cenchrus ciliaris* (Buffel grass), indigenous to Arabian Peninsula, has been identified as one of the most tolerant forage grass to drought during vegetative and reproductive stages in the field and is currently being used as a fodder under experimental conditions in the UAE and in parts of Australia, South Africa and Pakistan under commercial conditions. In addition, this species is a highly nutritious grass and is valued for its high yields of palatable forage and intermittent grazing during droughty periods in the tropics in addition to its low cost of establishment, tolerance to drought conditions and crop pests, and its ability to withstand heavy grazing and trampling by livestock.

Developed seeds would be ready for germination following dispersal or would have a kind of dormancy that may delay their germination until the arrival of the favorable season for seedling survival. The main causes of seed dormancy are (1) a physiological inhibiting mechanism of germination in the embryo (= physiological dormancy), (2) seed coat impermeable to water (= physical dormancy), or (3) underdeveloped embryo (= morphological dormancy). Seed dormancy can influence patterns of plant distribution, recruitment dynamics and persistence in the plant community. The advantages of dormancy are to enable seeds to accumulate in the soil seed bank and protect plants from expanding their entire reproductive output at a given time. On the other hand, maintenance of seed dormancy when conditions are optimal for germination can be disadvantage, as the seeds are exposed to lethal environmental factors such as granivory and extreme temperature for long time.

Seed dormancy is generally due to several factors either internal and/or external. The internal factors include the impermeability of the teguments to water or oxygen, presence of some inhibitors in the seed coat, changes in internal hormonal balance or physiological immaturation of embryos. Several factors have been reported to overcome seed dormancy including storage under different conditions, removal of bracts or dispersal organs enclosing the seeds, presoaking the seeds in warm water, and treatment of seeds with different dormancy regulating substances. Therefore studies are needed to assess dormancy level through the estimation of final germination percentage and germination rate as a result of (1) presence and absence of floral parts around the seeds (i.e., dispersal units or spikelets vs. naked seeds), (2) presoaking the seeds in water for different durations and different temperatures, (3) seed storage for long time periods, (4) treatment of the seeds with different dormancy regulating substances, and other related issues.

1.2 *Water management challenges*

The following major water management issues are:

1.2.1 *Water use: policy, planning and regulation*

- Reduction in quantity and quality of groundwater through over-abstraction, resulting in salinization of land, reduction in crop yields and abandonment of farms.
- Lack of farm management leading to over-irrigation and drainage problems
- Unplanned development in the farming & forestry sectors
- Poor performance of the Forestry Sector due to insufficient water and poor water quality
- Little or no effort to manage the demands for water in agriculture sector
- Lack of recognition of the true economic cost of water when assigning its use
- Uncontrolled and un-regulated well drilling, leading to dry wells and wasted resources

1.2.2 *Protection, conservation and monitoring of water resources*

- Lack of a coordinated Emirate-wide water resources monitoring network and programme
- Groundwater pollution due to fertilizer use
- Lack of groundwater protection policies, e.g. no protection zones for municipal wellfields that still produce water of drinking quality, for example

- Lack of inventories on sources and demands
- Lack of qualified, technical, on-site supervision for drilling water wells
- Poor monitoring and data collection during drilling
- Insufficient water resources monitoring
- General waste of water and leakages
- Poor practices of water data and information management
- Non-availability or poor access to water resources information and data, and lack of a central, Emirate-wide database to hold and analyse water resources data and information
- No well inventory, poor data collection when drilling wells

1.2.3 *Coordination of groundwater exploration and assessment*

- Need for expansion of groundwater exploration programmes, especially for deeper aquifer potential
- Lack of coordination and collaboration between existing groundwater exploration and assessment programmes.

1.2.4 *Local, regional & International cooperation and collaboration*

- Little or no technical cooperation with neighbouring Emirates and countries, especially on developments on or near to the international boundaries

1.2.5 *Strategic emergency water resources*

- No developed strategic reserve of potable quality water in case of emergency (current reserve for less than 2 days)

Common to the solution of most of the issues and problems listed above is the requirement for the establishment of a central, independent authority for Water Management in Abu Dhabi Emirate. Up until recently, responsibility for managing water sources and water use was divided between several organizations and agencies. This fragmented arrangement is unsatisfactory for effective water management and results in lack of coordination and collaboration between some of the bodies, duplication of efforts, non-assignment of responsibility for some very important aspects of water management, wasted funds and a general lack of accountability for some organizations current practices which go against the principle of sustainable water resources management and rationalization of water use.

1.3 *Groundwater recharge*

Groundwater recharge projects are considered necessary for meeting present and future demands on limited water resources. Continued growth in arid regions will require the development of a sustainable water supply, which will no doubt include improvement of our ability to store and utilize reclaimed wastewater.

Groundwater recharge using either relatively clean water or treated wastewater involves the infiltration of surplus or impaired quality water from the ground surface to underground water storage aquifers. Water is captured and stored underground in times of surplus for use in times of deficit. Underground storage avoids some of the disadvantages of conventional surface water reservoirs, such as evaporation, potential exposure to contaminants, and large capital investment become apparent.

Particularly in arid lands, treated wastewater has gained acceptance as a water supply and as an augmentation to the more conventional surface and groundwater supplies. Infiltration land treatment systems, wherein the groundwater is recharged with pretreated wastewater effluent have been suggested as a viable treatment option in arid lands.

Studies have shown that the hydraulics of water recharge basins are affected by many factors such as soil type, surface non-homogeneity, the formation of a surface clogging layer, application times for wetting/drying cycles, water quality, and climatic conditions. Clearly maintenance of a reasonable hydraulic loading capacity is of primary importance to successful soil aquifer treatment technique. Thus, a balanced approach to surface clogging layer and soil type selection for obtaining appropriate levels of water quality as well as rate of recharge must be achieved. Therefore, it is important to understand the movement of recharged water through unsaturated soils and to understand the role of unsaturated soils in the treatment process when the source of recharged water is not treated.

1.4 *Groundwater contamination and management*

Issues of concern to groundwater contamination and management are:

a. Development of innovative barrier systems for containment of hazardous and non-hazardous wastes in arid lands;
b. Development of comparative information to enable more systematic site selections, which will, in turn, make maximum usage of the natural attenuation capacity of given environmental settings to prevent groundwater contamination;
c. Environmental authorities should increase its attention to prevention of groundwater contamination and should provide programs to achieve a greater balance between prevention and remediation activities; and
d. Development of Best Management Practices (BMP) for usage of groundwater resources.

1.5 *Monitoring systems*

The use of low permeability lining materials has proven an effective method for containment of many types of waste leachates and could benefit from the use of an *in-situ* detection technique to monitor the performance of these liners. Current monitoring procedures rely mainly upon monitoring *wells*, *lysimeters*, and *leachate under-drains*. Wells are the most common means of monitoring the ground water contamination. This approach tends to be expensive and time consuming to implement. Timely detection of contaminant plume is obviously dependent on the initial layout and a number of monitoring wells. Unfortunately, wells can sample only a small volume of the aquifer. If samples collected from wells are not representative of the area or conditions for which they are intended, misleading and erroneous conclusions may result. Experience has shown that by the time a contaminant becomes detected in a monitoring well, a substantial volume of the surrounding soil and groundwater has already been polluted. In addition in arid lands, it is difficult to impossible to collect such required volume of water sample because of the high hydraulic conductivity of such soils. The risk of drilling wells and exploratory holes in unknown hazardous waste sites can be substantial. As the number of holes needed to define a problem area increases, so does the possibility of puncturing buried containers. Toxic fumes and liquids may be released. Explosions and fire may occur in extreme cases.

The limitations associated with present monitoring techniques underscore the need for an alternate approach. Undeniable, early detection and characterization of subsurface contamination can minimize its negative impact. Therefore, there is an urgent need for the development of a field diagnostic technique that allows a rapid determination of the extent of pollutants present in subsurface soils. The developed method should assist in locating a leak in the impounding boundary so that a corrective action can be taken to alleviate the problem. It should also be adaptable to a wide range of chemicals, as opposed to being ion specific.

There is an urgent need for the development of non-invasive pollutant detection systems in vadose zone for both organic and in-organic pollutants. Systems such as electrical polarizations, time domain reflectometry, and fiber optics chemical sensors are highly recommended and need further research studies.

1.6 *Solid waste management and utilization*

Due to the advancement in the understanding of landfill behavior and decomposition processes of municipal solid wastes, there has been strong thrust for upgrading existing landfill technology from storage/containment concept to process-based approach, i.e., bioreactor landfills. Bioreactor landfills allow a more active landfill management that recognizes the understanding of biological, chemical, and physical processes involved in a landfill environment. Engineered bioreactor landfill sites can provide:

a. a more controlled means by which the society can reduce the emission of global warming landfill gas;
b. immediate improvement to the surrounding local environment;
c. accelerated waste biodegradation, a means to enhance landfill gas generation rates; and
d. overall reduction in landfilling operation and maintenance costs.

Therefore, to enhance ones ability in applying such technology, there is a need to further understand the physical-chemical-biological processes that generally take place during the operation of landfills.

1.7 *Treatment of contaminated soils*

The use of microbial enhanced oil recovery concept to treat soils contaminated by hydrocarbons is a promising technology. The technology has been proven in oil industry for enhancement of oil recovery from underground reservoirs. The technology has been proven in oil industry for enhancement of oil recovery from underground reservoirs.

1.8 *Waste water treatment*

The development of petroleum and petrochemical industry in the Arabian Gulf region has been growing rapidly. In the last 50 years, exploration, drilling, extraction, refining and chemical engineering activities of oil and gas industry have all become an essential component in the economy of many of the Arabian Gulf countries. This speedy development has resulted in many changes such as landscape, economy, human development and interactions with other regions of the world as well as having impacts upon the environment and society. In fact "one of the most serious challenges facing the modern Middle East is the protection of its environment and the need to balance sustainable development with environmental security."

Water pollution is a serious concern in the UAE. This critical problem is made even more serious with the fact that water is very scarce in the region. Like most industries, oil refineries generate enormous quantities of wastewater. Such wastewater may contain several pollutants including phenols. Because of its toxicity, phenol became a wastewater quality parameter that the regulators closely look at in the effluents of heavy industry such as refineries.

Several studies have been reported in the literature on the use of biodegradation for removal of phenols from wastewater. In many of these studies, Pseudomonas putida (Pp) strain-type bacteria have been found to be effective for the degradation of phenols. The efficiency of biological methods employed for the treatment of phenol-contaminated wastewater varied with operating process variables such as pH, temperature, dissolved oxygen, and nutrients. It was further found that anaerobic degradation is typically more effective in removing phenol than degradation under aerobic conditions.

Most of the studies conducted on the removal of phenol from wastewater were limited to laboratory conditions. Transfer of laboratory results to filed conditions may not be straightforward, especially when processed wastewater is generated at condensate crude oil refineries. These refineries commonly receive crude oil from various sources, leading to fluctuations in waste characteristics. Studies to investigate the efficiency of treatment processes used in oil industry in reducing phenol concentration are needed.

1.9 *Environmental impact assessment*

Issues of concern to environmental impact assessment are:

a. A key issue that may influence the environmental impact assessment findings and subsequent interpretation is that the subsurface environment is typically characterized by very non-uniform conditions. Accordingly, consideration of such non-uniformity must be included in project evaluation and in interpretation of anticipated impacts.

b. Another issue of concern is the difficulty of taking a consistent approach for impact interpretation given the rapidly changing regulatory programs.
c. Management of soil and groundwater quality represents a new field, and considerable changes are occurring relative to the development and implementation of pertinent legislation, standards and regulations.
d. Environmental indicators and/or environmental indices can be useful tools in preparing a description of the environmental setting for a proposed project. There is a need for the development of indices for the impact of vadose zone on groundwater pollution potential.
e. Because of the need to investigate information from a number of substantive areas in the environmental impact assessment process, and because of the relative newness of the field in environmental management when compared with non traditional disciplines such as biology, chemistry and environmental engineering, there is a need for practitioners who work on the planning, conduction, and review of environmental impact studies to receive appropriate training.
f. Planning and implementation of research programs focused on answering specific impact related questions.

1.10 *Risk management of polluted sites*

Any successful approach to risk management must address a multiplicity of technical and regulatory issues. Such complexities demand the detailed attention of a coordinated, multi-disciplinary project team and integration of issues in a way perhaps unprecedented in other engineering and scientific endeavors.

Inherent uncertainties in site conditions, geochemical data, regulatory criteria, and technology performance must be recognized in formulating a risk management approach. Regardless of the level of efforts expended, uncertainty cannot be eliminated and can only be managed through an approach that recognizes this fact, strives to understand the certainties, and establishes and implements and adaptive methodology.

Regulatory attitudes exert considerable influence on the design aspects of waste containment facilities. The assessment of failure of the facility to function in a manner designed to ensure protection of public health requires one to properly appreciate what constitutes a health threat. However, it is not immediately clear that the controls needed to establish safe protection of public health are well founded. A good working knowledge of the various interactions occurring in the substrate during pollutant transport is required if one seeks to assess the fate of pollutants.

1.11 *Sulfur utilizations in construction industry*

A dramatic change has occurred in the global sulfur industry over the last decade. Sulfur consumption is dominated by a number of large-scale uses, with several complex interactions that complicate the future outlook for demand. Thus, sulfur is used to produce sulfuric acid, one of the world's largest volume chemical commodities, but this use competes with sulfuric acid recovered from ore smelting. Uses of sulfuric acid include processing phosphate rock and ore leaching in metallurgical processing (a growing market). There are a number of other uses today for sulfur, including in pigments, pesticides, and rubber vulcanization and as an agricultural nutrient.

In general, existing uses for sulfur are relatively mature, and offer limited opportunities to consume significant new supplies. Currently sulfur is in net surplus on a global basis, and with environmental regulations mandating ever greater sulfur recovery from petroleum and gas processing. The outlook is clear; there will be substantial and growing surpluses in global sulfur supply for the foreseeable future. Sulfur prices are likely to be under pressure, and producers could face substantial and growing disposal fees.

Sulfur in its elemental form is recognized as an important ingredient in several agronomic applications. These include the following:

1. an essential plant nutrient;
2. an active agent for increasing crop stress resistance;

3. environmentally benign pesticide; and
4. an efficient soil amendment aid to alleviate alkalinity.

Sulfur as an essential plant nutrient has received little scientific attention. This is explained by the facts that sulfur was obviously in sufficient supply from the atmosphere, from soil and as a by-product in mineral fertilizers. However, the use of highly concentrated fertilizers containing little or no sulfur has drastically reduced the amount of sulfur supplied to soils. Recent studies have shown that adding sulfur to soil increased crop yield, increased drought tolerance and increased nitrogen efficiency and phosphorus uptake.

All of these applications are important to the national agricultural drive in the Gulf Cooperation Council (GCC) countries. However, the fourth application, which is alkalinity amendment, is of particular significance. Soil alkalinity, in the GCC, demands the use of an acidifying agent to achieve the required neutrality. Sulfur is the major component to achieve this endeavor. Elemental sulfur is microbially oxidized to sulfuric acid, which then reacts with calcium carbonate to form gypsum. This oxidation process is highly dependent on soil moisture, temperature, microbial activity, and the size of the elemental sulfur grain. Particle size is perhaps the most critical factor from an application and product point of view.

In general, elemental sulfur granules, in their original size (250 μm), oxidize at a very slow rate. These granules are relatively large and present soil microbes with a small specific surface area for conversion. By breaking the elemental sulfur granule into smaller sizes (45 μm), the surface area is increased which in turn increases the rate of microbial oxidation and conversion of elemental sulfur to sulfate.

The phosphate fertilizer industry is the largest consumer of sulfur, primarily as a consumer of sulfuric acid used in phosphate rock processing. Sulfur itself is also a plant nutrient, mostly supplied as ammonium sulfate and potassium sulfate, and elemental sulfur is a traditional fungicide. Agriculture is thus a very large, but modestly growing outlet for sulfur.

Global sulfur demand has been relatively stagnant at about 57 million metric tons per year over the last decade. Based on new regulations limiting sulfur content in diesel and gasoline, the current small global surplus in sulfur supply is projected to reach between 6 and 12 million metric tons by 2011, or between 10 to 20 percent of demand. This projected surplus represents obvious challenges to existing producers, potentially leading to drastically reduced sulfur prices, and even the possibility of costs to producers for disposal of the surplus in some regions. On the other hand, the surplus may also represent opportunities for new uses of sulfur, driven by these very same reductions in sulfur price that can make new uses more economically feasible and attractive.

Even with relatively small surpluses, the oil and gas industry has already experienced strains on sulfur storage facilities. Sulfur is being stored on site in block, granular or palletized form, or molten, (very expensively) in rail tank cars because there is insufficient storage capacity to handle sulfur generation at refineries and gas processing plants.

One potential new market is ***sulfur solidified concrete***. This is a thermoplastic composite of mineral aggregates bound together with chemically modified sulfur. The product is more durable than Portland cement. In addition, from the environment viewpoint, there may be monetizable benefits in reducing greenhouse gas emissions that would enhance the attraction of sulfur solidified cement applications.

The main advantage of sulfur concrete is its use as a highly durable replacement for construction materials, especially Portland cement concrete, in locations within industrial plants or other locations where acid and salt environments result in premature deterioration and failure of Portland cement concrete.

There are several advantages in using sulfur concrete for construction in areas exposed to highly corrosive acids. While ultimate life or durability of sulfur concrete has not been completely established in many end use applications, enough evidence of its corrosion resistance and durability has been accumulated to show that it has several times the life of other construction materials now being used in corrosive environments. Other advantages of sulfur concrete are its fast setting time and rapid gain of high strength. Since it achieves most of its mechanical strength in less than a day, forms can be removed and the sulfur concrete placed in service without a long curing period.

The handling, mixing, and use of sulfur concrete can be accomplished with proper concern for product safety. As with any other construction material, certain measures must be taken with sulfur concrete to insure safe handling in its preparation and use. Sulfur concrete should be produced within its recommended mixing temperature range (127–141 deg. C) to minimize emissions. Adequate ventilation during construction operations and normal precautions for handling hot fluid materials (proper protective clothing, eye protection, gloves, and hard hats) should be observed.

Sulfur solidified concrete has potential use in a wide range of applications such as:

1. pre-casting activities (sewer pipe; railway ties; highway barriers; a range of agriculture products; offshore drilling platforms; construction blocks, slabs, and other building components);
2. extruding (bricks and paving stones; curbs and gutters; roof tiles); and
3. cast-in-place (bridge decks; marine installations; ship hulls and structures exposed to marine environments; drilling platforms; food processing plants; agricultural applications, including barns and effluent systems; sewage treatment plants; acid plants-drainage canals, sumps, tanks, flooring, walls and beams; fertilizer plants; foundations).

To appreciate the market size for such product in the United Arab Emirates (UAE), one briefly reviews the size and operation of cement manufacture. In UAE there are nine major cement manufacturing plants, eight producing Portland cement and one producing White Cement. The production capacity of Portland cement plants in 2000 was over 10 million tones per year (Arab Union for Cement and Building Materials, 2003) and that of the White Cement plant was about 450,000 tones per year. In 2000, these plants produced around 6,900,000 tones of cement generating over one-billon UAE Dhs (about 271 million USD) in income. Furthermore, the net profit of cement producing plants in UAE has increased from 73.3 million UAE Dhs in year 2002 to 278.5 million UAE Dhs in year 2003.

Sulfur concrete is a thermoplastic construction material, of mineral aggregates and modified sulfur as binder. Elemental sulfur has a low smelting point (120 deg. C) and appears in two solid allotropic forms, separated by considerable contraction. Hence, there is a need for modification to delay the transition. There are two main ways of modifying sulfur, both of them trying to control sulfur crystallization (unmodified sulfur crystals are brittle and subjected to failure under thermal cycles). The first one is "Plastification", where a certain chemical substance (a polymer) is added to sulfur in order to inhibit the transformation to orthorhombic structure; as a result of the chemical reaction with the substance, after cooling sulfur remains in monoclinic state, with a amorphous fraction also. Several substances have been tried for this; the most common are dicyclopentadiene, or a combination of dicyclopentadiene, cyclopentadiene and dipentene.

The second method is "controlled crystallization", where, although sulfur is allowed to reach the orthorhombic state, crystals are small and with a regular and ordered structure. To obtain this, sulfur is modified with an olefin polymer (such as RP220 or RP020 by Exxon Chemical, or Escopol) and a physical stabilizer is also added (for example, fly ash, or other fine substance). In both ways, the issue is controlling sulfur crystallization, either chemically or physically. Depending on the ultimate use of the produced sulfur concrete, one chooses the method of treatment. Furthermore, with proper content of physical stabilizers, sulfur concrete tends to be self-extinguishing. Also it can be manufactured to be fire resistant, and/or inhibit the formation of SO_2 when heated with proper content of polymeric materials. Therefore, more research studies are needed in this area to develop the technology on large scale.

1.12 *Sustainable development*

Sustainable development has become public concern in recent years. In order to achieve sustainable development, cooperation at a global level is important to control the environmental burden caused by human activities to a degree that can be tolerated by the planet.

The three major players for achieving sustainable cities are economy, environment and society. Instead of working on the improvement of each of the three spheres separately, a result-oriented approach is to work for interlinking the economy, environment and society. Actions to improve conditions in a sustainable community take these interlinks into consideration. Sustainable environment depends on sustainable development, which in turn linked to sustainable construction practices. Engineers, Architects and Construction Managers can improve the energy and resource efficiency of commercial and residential buildings by the use of practices for water conservation, better indoor quality, use of day light, energy efficiency for which we need to impart relevant technical education and skill training. Without the emphasis of these aspects in technical education and skill training the approach towards sustainable cities is rather going to be incomplete. First starting from awareness among higher secondary school classes and later inclusion of these aspects into engineering and technical education and in skill training in vocational institutes will provide the desired results.

There is a need to introduce and include sustainable building concepts in engineering and technical education for having sustainable cities. Some introduction should also be given in general and science education, so that city people who are going to use a house or a building can ask/demand for better indoor quality, energy efficiency, etc. Further research in sustainable cities and sustainable development is needed for the better and efficient use of our resources keeping in mind healthier environment for urban population and future generations. The three areas of economy, environment and society needs to be interlinked for a purpose oriented approach. Use of local/indigenous building materials needs to be encouraged. And finally, research in local low cost environmental friendly building material to be carried out.

1.13 *Global warming*

The threat of climate change caused by the emission of anthropogenic carbon has encouraged the international community to develop agreements such as the United Nations Framework Convention on Climate Change and the Kyoto Protocol. Upon ratification of the Kyoto Protocol, Canada committed, through the 2008–2012 commitment period, to reducing its greenhouse gas (GHG) emissions to 570 MT eCO_2, from 1990 levels of 610 MT eCO_2. The solid waste sector in Canada generated 24 MT eCO_2 in 2000, 23 MT eCO_2 of which were produced by landfill gas (LFG). The transport of waste likely generated a further 740 kT eCO_2 that are not accounted for in the waste sector by the National Greenhouse Gas Inventory. It is likely that waste transport emissions will increase as fewer landfills are sited and further source separation of wastes occurs. The benefits of source reduction, recycling, LFG capture for energy recovery, composting, anaerobic digestion and incineration are parameters that need to be studied.

In its voluntary action plan for fiscal year 2010, Japan's oil refining industry established a goal of "reducing unit energy consumption in refineries by 10%." Owing to efforts that aimed to achieve the goal, it has produced results that exceed the goal. Cumulative energy conservation is estimated as approximately 3 million KL. When calculated from as far back as 1973, the year of the first oil shock, energy conservation effect amounts to a total of approximately 10 million KL. In this way, the oil industry worked to reduce CO_2 emissions through the steady implementation of energy conservation measures in refineries. At this rate, with continued efforts it should be sufficiently possible to achieve the fiscal year 2010 targets.

Japan's Kyoto Protocol Target Attainment Plan states that, in order to fulfill the commitment to achieve a 6% reduction over the value of 1990 which Japan pledged in the Kyoto Protocol, Japan must implement measures to achieve an additional emission reduction corresponding to -12% (approx. 148 million tons-CO_2 equivalent) as well as policies for their promotion in addition to those measures and policies that it are currently being implemented. The voluntary action plan of the Keidanren adopted by 34 major industries calls for a reduction in CO_2 emissions to below 0% from fiscal year 1990 to 2010. In these ways, with the steady implementation of voluntary action plans in the industrial sector, achievement of the targets looks promising. In regards the issue of CO_2 emission control in the private transportation sector, the shift to sulfur-free gasoline, diesel fuel, and other transportation fuels is expected to produce effective results in the sector.

Efforts related to reducing CO_2 emissions through the supply of environmentally compatible petroleum products are welcome. It is our hope that industrial countries will effectively introduce measures to the rationalization of energy use and the production of environmentally compatible products by oil factories throughout the world in order to achieve the current goal of reducing greenhouse gases.

1.14 *Oil spill management*

Oil spills in the marine environment raise a major concern among platform management, government authorities and the public. One of the key elements for an efficient spill control is the preparedness for an adequate and prompt response. Studies for developing framework for formulating an oil spill emergency response plan are needed.

2 PROPOSALS FOR ACTIONS

Certainly, ***individual initiatives*** are very important if the Gulf Region is to become a strong and recognizable entity in the development of Sustainable GCC environment and water resources. These initiatives can be better ***vectorized*** by the creation of identifiable units which can act to the common advantage of the environment and water resources communities in this region. These can include the creation of:

1. The JCCP-GCC Working Group for EWR series of symposia; and
2. The JCCP-GCC Research Fund.

Groundwater management

Arid Land Hydrogeology: In Search of a Solution to a Threatened Resource – Mohamed(ed.)
© 2006 Taylor & Francis Group, London, ISBN 0 415 41127 0

Groundwater resources: Development & management in the Emirate of Abu Dhabi, United Arab Emirates

M.C. Brook, H. Al Houqani, T. Darawsha, M. Al Alawneh & S. Achary
Environment Agency, Abu Dhabi, UAE

ABSTRACT: With Abu Dhabi Emirate's current annual water use twenty six times larger than its annually renewable natural water resources, there is an urgent need for implementation of programmes and projects to improve water management and rationalize water use in all sectors. An analysis of all water produced shows that the vast majority (79%) is groundwater abstracted from boreholes and shallow hand dug wells and given declining water levels, and a general deterioration in groundwater quality, the protection and conservation of all groundwater, irrespective of its quality, is of vital importance. This paper describes the groundwater resources and water management issues and challenges associated with the various groundwater sector users in Abu Dhabi Emirate. A three year programme of numerous collaborative projects and activities with some of the twenty or so other stakeholders involved with either producing or using water in the Emirate commenced in 2002 and has resulted in the general consensus for the requirement for one body to be provided the responsibility for water resources management. In February, 2005, the mandate for groundwater management, including exploration, assessment and monitoring was given to the Environment Agency Abu Dhabi (EAD). The activities of EAD, which are described herein, now largely focus on the priorities of regulation of water use, protection and monitoring of Water Resources, developing a central water database and managing strategic water reserves.

1 INTRODUCTION

Abu Dhabi Emirate, one of the seven Emirates which comprise the United Arab Emirates (UAE), occupies an area of 67,340 km^2, or about 80% of the total area of UAE (Figure 1). The Emirate has an arid climate with less than 100 mm/yr average rainfall, a very high evaporation rate (2–3 m/yr), a low groundwater recharge rate ($<$4% of total annual water used) and no reliable, perennial surface water resources. Furthermore, it is a downstream water user and shares trans-boundary water resources along common borders with Saudi Arabia and the Sultanate of Oman, 350 km and 280 km in length respectively.

The Emirate occurs in the subtropical arid climatic zone and is exposed to oceanic effects of the Arabian Gulf and Indian Ocean. Rainfall is erratic and unreliable. Orographic effects are clearly seen on the rainfall distribution. The Al Hajar mountains in neighbouring Oman, which reach elevations in excess of 2000 m.a.m.s.l, generate high rainfall incidents, especially in the winter months, which provide for the runoff to wadis which cross over the boarder into Abu Dhabi Emirate. Within Abu Dhabi, this high elevation rain occurs only at Jebel Hafit, which, at an elevation of 1163 m.a.m.s.l, is the highest point in Abu Dhabi Emirate and the only high mountain massif within the Emirate.

Mean annual rainfall within Abu Dhabi Emirate varies from 46 mm at Jebel Dhana in the Western Region to 119 mm at Al Wigan, south of Al Ain, in the Eastern Region. The mean annual rainfall at Al Ain 1971–1994 is 96.4 mm.with a maximum of 303 mm/yr. The mean annual precipitation for Abu Dhabi Island is 87 mm, with a maximum of 227 mm/yr.

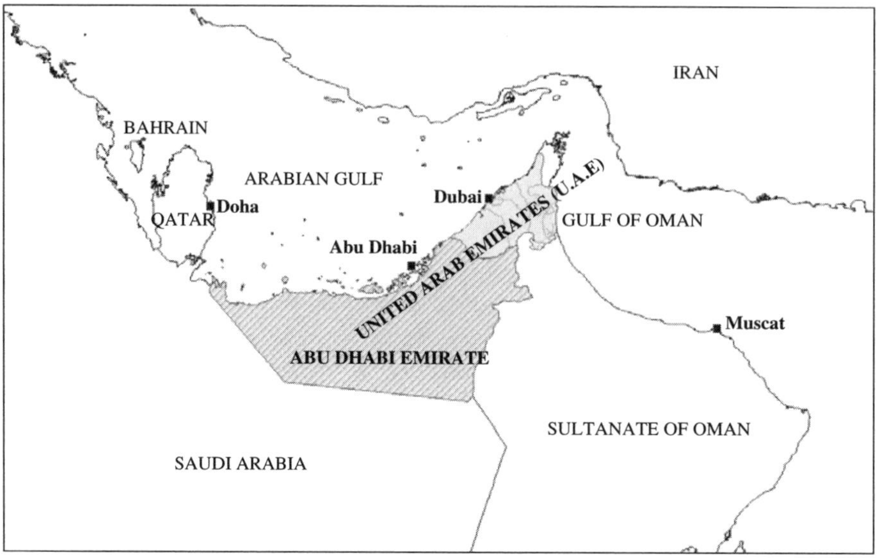

Figure 1. Location of Abu Dhabi Emirate, United Arab Emirates.

Groundwater, albeit mostly brackish and saline in quality, still provides around 80% (ERWDA, 2003) of all water used in the Emirate. No natural, perennial surface water resources exist within the Emirate, apart from the spring at Ain Al Fayda and it is therefore important to understand the various hydrologic processes and hydrogeologic settings which control the extremely valuable groundwater resources.

Historically, all the Emirate's water requirements were met solely from groundwater obtained from shallow hand dug wells and the traditional Falaj system, comprising man-made channels used to collect groundwater, spring water and surface water and transport it, by gravity, to a demand area. Since the entire Emirate's Aflaj are now dry, a system of borehole support has been developed over the last 5–10 years.

Over the last two to three decades, however, rapid economic development, coupled with sharp population increases and the development of a large agricultural sector, substantially supported from Government subsidies, has meant an increasing reliance on unconventional water resources, such as desalination, and also the development of alternative conventional water supply measures, such as recharge dams, storage dams, recharge wells, interception of groundwater losses, re-use of wastewater and water transfers.

Figure 2 shows a breakdown of water sources and consumption for the year 2003 and Table 1 shows detailed consumption for the Eastern and Western Regions. Figure 3 shows the location of sources and users of water. Irrigation in the agriculture, forestry and amenity plantation sectors accounts for a massive 82% of the total, the remainder is taken up with domestic and industrial consumption, both of which is supplied in bulk by the Abu Dhabi Water and Electricity Authority (ADWEA).

2 GROUNDWATER RESOURCES

2.1 *Groundwater systems*

These systems are controlled by recharge processes, the geology of the host rocks, and the residence time of groundwater and discharge processes. The resultant groundwater quality is largely influenced by groundwater residence time and type of recharge process, and in more recent times, by anthropogenic activity and its availability by aquifer type and surrounding development. Table 2 gives

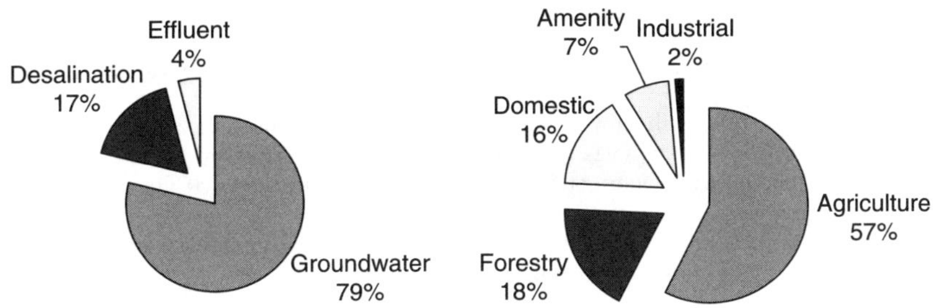

Figure 2. Sources and users of water in Abu Dhabi Emirate, 2003.

Table 1. 2003 Water consumption.

	Eastern Regin (mm^3)	%	Western Region (mm^3)	%	Total (mm^3)	Total %	Change Since 2002%
Domestic	136.87	9.16	385.13	20.41	522.00	15.44	+18
Industry	15.21	1.02	42.79	2.27	58.00	1.72	0
Agriculture	1109.07	74.19	840.29	44.54	1949.36	57.64	−1
Forestry	122.85	8.22	484.45	25.68	607.30	17.96	+18
Amenity	111.00	7.42	134.04	7.10	245.04	7.25	−5
Total	1495	10	1886.70	100	3381.70	100	+5

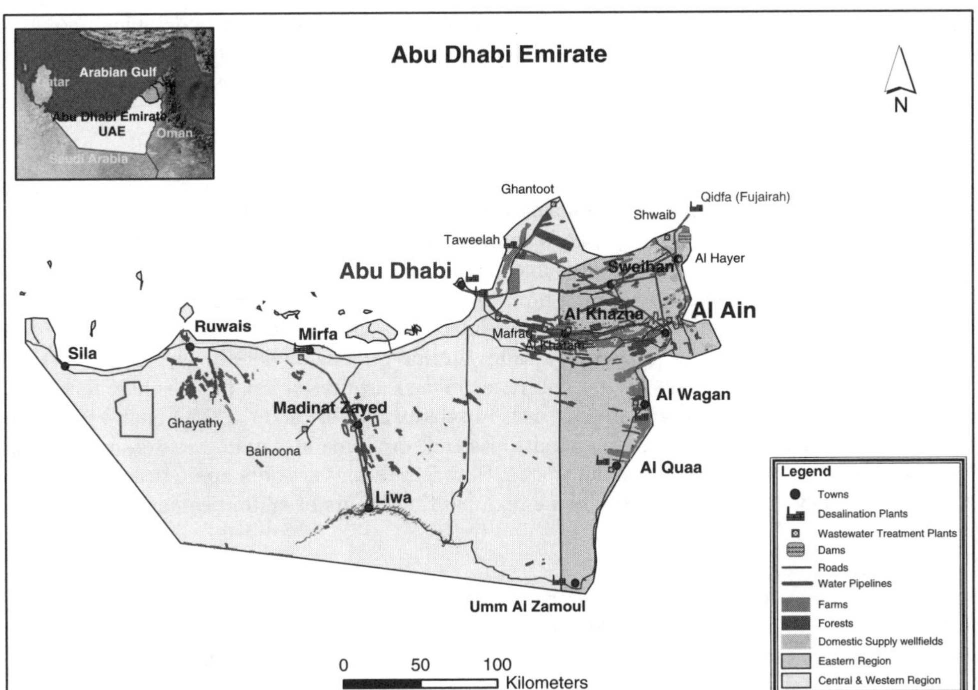

Figure 3. Sources and users of water in Abu Dhabi Emirate.

the main characteristics of the three flow types: local, intermediate and regional. Overall groundwater movement is generally from East to West for all three flow types, although, North of the Liwa crescent, a groundwater high allows flows to the south and across the border to Saudi Arabia. Flow times from recharge zones in the East to the sabkha discharge zones along the Gulf coastline

Table 2. Groundwater flow systems within Abu Dhabi Emirate.

System type	Main physical & development characteristics	Main hydrochemical characteristics
Local	Occurs as springs, shallow hand dug wells, aflaj and shallow boreholes within surficial gravel and alluvium aquifers. Short groundwater residence time in active recharge areas, rapid hydrological cycle. Limited to Eastern region, close to Oman boarder	Low salinity & temperature and close to ambient air temperature. Groundwater of Bicarbonate (HCO_3) type, indicative of active recharge e.g. Al Jaww plain, Gashaba, Shuwaib areas
Intermediate	Inland sabkhas are main discharge areas. Groundwater contained in relatively thin sand aquifers, low groundwater velocities with moderate residence times	Generally brackish to saline and of Sulphate (SO_4) type. Hypersaline at discharge areas i.e. Sabkha. No or little active recharge, most of Western region has this system, although Ain bu Sukhanah spring at Ayn Al Fayda also belongs to this system
Regional	This slow moving, long residence groundwater system moves towards the North West and the Gulf and also to the South West into Saudi Arabia where discharge areas are low lying sabkhas	Discharge areas have waters of high temperature and are highly mineralized. Sabkhas are hyper-saline. Residence times of up to 15,000 years produce Chloride water types

can take up to 15,000 years. This slow groundwater movement allows for considerable dissolution of salts in the groundwater and the longer the residence time, the higher the salt content; hyper-saline waters in excess of 200,000 mg/l are found along the Abu Dhabi coastline.

2.2 *Aquifer types*

Around 80% of the territory comprises Quaternary sand and sand with gravel aquifers. The Eastern Region main aquifers are Quaternary sands and gravels (underlain by the very productive Upper Fars Formation which continues eastwards into neighbouring Oman), the Lower Fars Formation in the south eastern, Umm Az Zamoul area, limestone bedrock units (Dammam and Simsima) and discontinuous carbonates within the tectonically effected hydrogeology north of Al Ain. The Western Region largely comprises the Quaternary sand aquifer directly underlain by the Lower Fars Formation as a basal unit, but unlike its occurrence in the Eastern Region, the Fars here represents a regional aquiclude. Also present are thin coastal sabkha aquifers and the Baynunah Formation, comprising continental Upper Miocene Sandstones and conglomerates with gypsiferous cap rocks that form numerous, low lying mesas in the area. Both Formations are poor aquifers and are largely undeveloped. They are both underlain by the regional Lower Fars basal unit aquiclude.

2.2.1 *Unconsolidated aquifers*

These are the most common and productive aquifers and comprise both recent dune sands and alluvial deposits of varying age. Collectively, the deposits comprise the surficial aquifers of Abu Dhabi Emirate or alternatively, the shallow (water table) aquifer, the top of which is defined by the water table. The following units of the shallow aquifer have been mapped by the Groundwater Assessment Project – GWAP (GTZ et al, 2005) and the overall thickness of the upper aquifer is shown in Figure 4.

- SA_L Quaternary aquifer/aquitard units directly underlain by the Lower Fars Formation as a basal unit (regional aquiclude)
- SA_U Quaternary sand and gravel aquifer underlain by the Upper Fars Formation as a basal unit
- SA_S Coastal and inland sabkhas

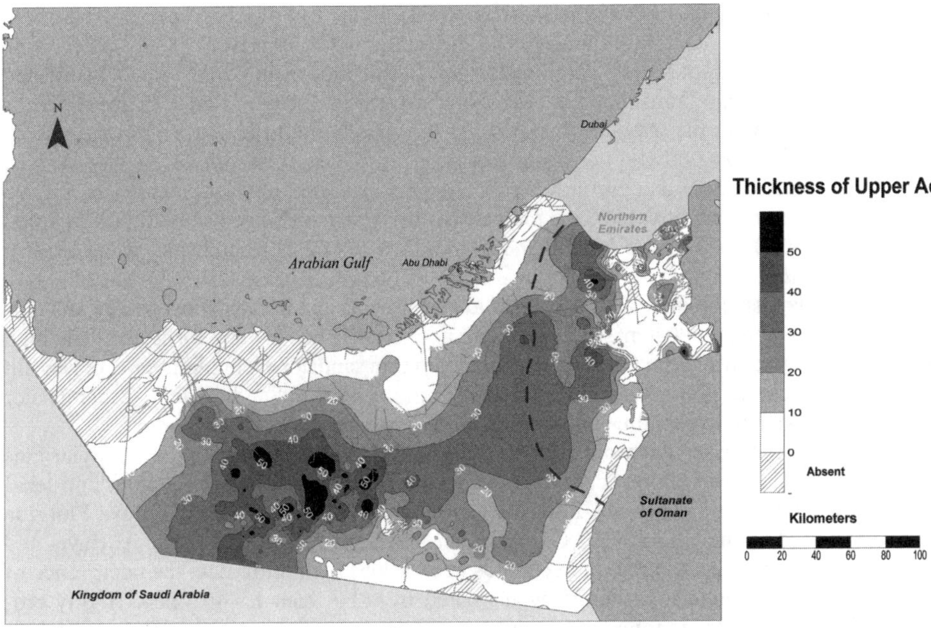

Figure 4. Saturated thickness of the upper aquifer (adapted from GTZ et al, 2005).

- SA_M Quaternary sand and gravel aquifer underlain by tectonically emplaced dark Marlstones and shales as main basal unit
- SA_J Quaternary sand and gravel aquifer east of Jebel Hafit (Al Jaww Plain) underlain by Upper Fars and Lower Fars Formations as basal unit.

Sabkhas are uneconomic aquifers and contain groundwater of hyper-salinity and brine quality in some places. Their hydrogeology and hydrochemistry is described in detail by Wood and Sanford (2002) and their potential for brine resources development by Czarnecki et al. (2000). The former concluded that rainfall is the dominant source of water whilst ascending terrestrial brine is the dominant source of contained solutes for the coastal sabkhas which comprise a 300 km long by 15 km (range 2–20 km) wide strip.

Much of the Emirate is covered with quaternary age (Holocene) aeolian sand deposits that comprise many different types of dunes, some of which occur as massive complexes. Dunes range in size from mega barchans found in the dune field directly south of the Liwa crescent where the average relative dune heights are 103 m, to small barchans south east of Baynunah which have an average relative height of less than 10 m (UAE University, 1993). The dunes are a common north–eastern extension of the well known sand sea "Ar Rub Al Khali" which lies mainly within Saudi Arabia. The dune sands aquifer, which is dominant in the Western Region, comprises mostly medium to very fine grained aeolian sands, sub-rounded to well rounded, which become progressively argillaceous with increasing depth. Individual sand grains are frosted, white to reddish-orange and composed of quartz, carbonate, and dark heavy and evaporate minerals. The upper parts of the aquifer are relatively clay and silt free and thus have moderate permeability and high porosity (NDC/USGS, 1993, USGS, 1996) and this permeable and productive zone is termed the Upper shallow aquifer which is underlain unconformably by the gravels and conglomerates in the Eastern Region and by a zone of lower permeability and productivity in the Western Region, termed the western aquitard (GTZ et al, 2005). This aquifer contains the fresh water basin north of Liwa Crescent (USGS/NDC, 1994) and is the beneficiary of the artificially recharged desalinated water introduced as a pilot ASR scheme in 2004. Another fresh water mound is also found in the dune sands of the Bu Hasa oil field (Rizk and Alsharhan, 2003). In the Liwa area, where 2,400 km^2 are underlain by fresh groundwater (Moreland, 1998), the dune sands comprise medium to very fine grained sand with silt composed of carbonates, quartz and heavy minerals. The unconfined Liwa aquifer has an average Transmissivity and specific yield of 300 m^2/d and 22% respectively. In the Liwa crescent area, the average thickness of the aquifer is 30 m and a total storage of 101,000 Mm^3, of which 16,000 Mm^3 is fresh, has been estimated (USGS/NDC, 1994). Whilst some, small degree of modern day groundwater recharge still occurs in the Liwa area, as proven by groundwater Tritium content values, the majority of the water recharged some 6,000 to 9,000 (Wood and Imes, 1995) years before present and so therefore the fresh groundwater water lense is largely fossil in nature.

The Baynunah deposit comprises poorly consolidated fluvial sand of late Miocene age and outcrops over an area of about 3000 Km^2. Sediments are horizontally bedded and form relatively high topography up to 60 m.a.m.s.l. The Formation can contain thin sandstone, conglomerate, clayey silt and gypsiferous sandstone beds. Sediment source is from the west in Saudi Arabia. Groundwater quality is high brackish to hypersaline and there is little development potential.

Alluvial aquifers comprise Quaternary sands and gravels. In the Eastern Region, alluvial fans coalesce into piedmont plains which occur on the edge of the Oman Al Hajar mountains and Jebal Hafit. Environmental tracers of Chloro-Fluoride Compounds (CFC) and Sulphur Hexa Fluoride (SF_6) have been used by the USGS/NDC Groundwater Research Programme in the Al Hayer and Al Jaaw Plain areas of the Al Ain region (Symonds et al, 2005) in order to date the occurrence of recharge events in the alluvials. Recharge events dated from between 12–45 years. Highly productive parts of the alluvium are found in coarser grained deposits which were laid down in palaeo channels which are now buried at depth (Fitterman et al 1991, Woodward & Menges, 1991, Rizk et al, 1998).The alluvium also extends beneath a large area of aeolian sand along the boarder with Oman, north of Al Ain. Mapped unit SA_U is the most extensive quaternary gravel sequence which is unconformably underlain by the Upper Fars Formation as the basal unit. This unit is tapped by most of the farm wells in the Al Khazna/Remah region where individual wells have been tested at rates of above 150 m^3/hr. Here the formation is 40–50 m thick and the groundwater salinity ranges from 1,500–10,000 mg/l.

The alluvial deposits, along with the sand dune aquifers are the most productive of the unconsolidated units in the Emirate and collectively, they form the "upper aquifer in the Emirate (Figures 4 & 5). The most productive wellfields are found north of the Liwa crescent and in the Al Khazna and Al Khader areas. Transmissivities range from $<$1–8,000 m^2/d with an average of 594 m^2/d.

The Upper Fars Formation is present throughout the Eastern Arabian shield and is a very productive aquifer. The Fars Formation has its type locality in Southern Iran where it outcrops. In neighbouring Sultanate of Oman, a $100 million groundwater fed drinking water supply scheme to 100,000 people has been developed in the locally named Al Masarrat Aquifer, to serve the Al Dhahirah region (Brook, 2001). The Upper Fars UAE equivalent underlies about 80% of the Eastern Region of the Emirate and comprises primarily conglomerates (moderate to highly productive) with inter bedded dolomitic marls, clay and siltstones. The dolomite can occur as meter thick, impervious beds which tend to compartmentalize the aquifers, resulting in multi aquifer layers with varying hydrochemistry. This phenomenon is seen within the units which occur in the Al Wigan/Al Quaa area south of Al Ain (Khalifa, 2004, Brook, 1994). The upper zone is in hydraulic connection with the quaternary shallow aquifer and permeability and porosity of the aquifer decreases with increasing depth. Deeper, thin conglomeritic lenses can also be productive down to depths of around 120 m below sea level. The Formation is found at its thickest in the Al Khazna area (400 m) and also west of Bida Bint Saud (300 m). In the Eastern Region, the Miocene Upper Fars Formation is differentiated from the Quaternary gravels that unconformably overlie it by the occurrence of more cemented conglomeritic layers which are intercalated with dolomitic marlstones and siltstones with mostly dolomitic matrices.

In the Western Region, the Upper Fars Formation consists mainly of marl and mudstone with interbedded thin sandy layers. These altered Tertiary deposits range in Transmissivity from 1 to 270 m^2/d (average 58 m^2/d) and have average well yields of 535 m^3/d (Bright & De Silva, 1998).

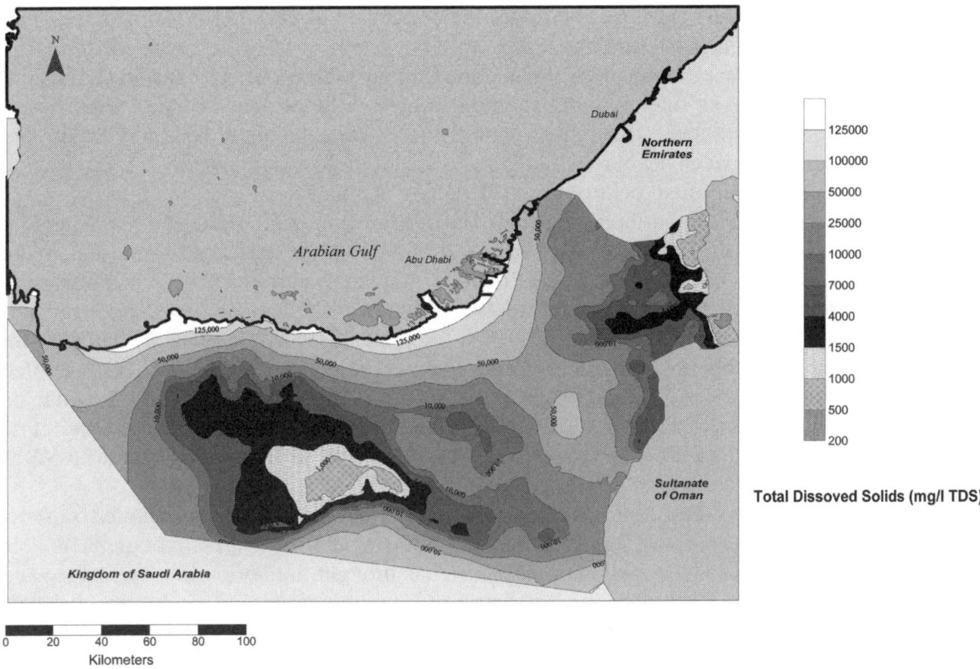

Figure 5. Groundwater salinity of the upper aquifer (adapted from GTZ el al, 2005).

The Lower Fars Formation occurs as thick (up to 650 m) early Miocene age mudstones and marly dolomites, intercalated with evaporites (gypsum and anhydrite) deposited in a shallow marine environment. It is present throughout the Emirate largely as an aquiclude and is unconformably overlain by Upper Fars and also the Western Region aquitard which comprises quaternary aeolian sands with frequent intercalations of inter-dunal sediments. In the Eastern region, the Lower Fars formed a sedimentary basin or trough that was filled with sediments of the Upper Fars Formation. Few wells have penetrated the Lower Fars and the top of the Formation has largely been mapped by use of vertical electrical soundings geophysical techniques.

2.2.2 *Bedrock and structural aquifers*

Bedrock aquifers occur throughout the Emirate and are largely carbonate deposits laid down in shallow marine seas. Their potential as aquifers has not yet been fully proven; the aquifers occur generally at significant depth and have not been explored or exploited anywhere near to the same extent as the unconsolidated aquifers described above. The Emirate can be divided into two structural regions as follows:

- Eastern – includes Eastern Region of the Emirate, underlain by eastern edge of the Arabian Shelf and the Oman Mountain Foredeep (Foreland Mobile Belt).
- Western – occupies 57,000 km^2 and occurs in western and central part of the Emirate and includes the relatively stable Arabian shelf Province and Rub Al Khali basin. Main structural feature is gentle, simple folding of Tertiary strata. The karstified and fractured nature of the strata produces multi-aquifer systems (Al Mardi, Al Aidrous. 1985).

The Asmari Formation occurs in both the Eastern and Western regions and has been relatively unexplored to date. In the Western Region, its equivalent occurs as an Oligo-Miocene clastic continental unit, comprising siliceous sand, sandstone and minor interbedded shale layers, whose upper contact is recognized by an overlying anhydrite bed at the base of the Lower Fars Formation (Imes et al, 1994). In the Eastern Region, the Formation has been mapped largely from seismic profiles and has an average thickness of about 200 m. In the region of Jebel Hafit, it is significantly folded and affected by thrust faulting. The ridges are dominated by coralliferous limestones with subordinate marls rich in fossils (Kirkham, 2004).

Kartsic limestone Formations occur throughout the Emirate but are only exposed in the Eastern Region where they occur as fractured and solution channeled limestones of late Cretaceous to Tertiary age within a north-south trending structural zone measuring about 25 km by 80 km that represents a transition between the buried, flat lying to slightly folded strata in the Western Region and the highly deformed, uplifted rocks of the Al Hajar Oman mountains. Rock outcrops are scarce but crystalline limestones can be found at Jebals Hafit, Muthaymimah, Malaqat, Oha, Masakin, Mohayer and at Qarn Tarab, Saba, Bida bint Saud and Mutarid in the northern structural domain. The various limestones occur along the axes of regional anticlines which have been mapped using data obtained from seismic surveys (Woodward 1990, Woodward & Jeelani, 1993) and structural analysis of borehole electrical images (Akbar, 1994, Akbar et al, 1995). Because the outcrops are associated with steeply plunging anticlines, the aerial extent of the productive aquifers is limited to shallow horizons near the exposures (Bright & De Silva, 1998) and the aquifers are found to be unproductive below depths of about 150 m below ground level. The Transmissivity and therefore the individual well yields within the limestones are highly variable, ranging from $5\,m^2/d$ to $8{,}700\,m^2/d$ and $200\,m^3/d$ to $6{,}000\,m^3/d$ respectively.

The Dammam and Rus limestone carbonate sequences have hardly been explored to date, largely because their occurrence at significant depth (>500 m below ground) although the double plunging anticline at Jebel Hafit and other anticlines has brought the Formations to surface and has allowed wells to penetrate the aquifer. Exposures of Dammam limestone can be seen at Jebals Oha and Hafit and also at smaller Qarns and Jebals along the eastern side of Al Jaww plain. The 15-well Mubazzarah wellfield has been developed at Jebal Hafit by Al Ain Municipality with a combined daily yield of about $4{,}600\,m^3/d$. Water supplied by the wellfield has been used for recreational purposes and for "greening" of the location. Because well fractures at depths between 100–200 m below ground intersect much deeper fractures that have their origin of up to 2,000 m below ground level, groundwater in some wells attain temperatures of greater than 50°C (Khalifa, 1997) and, because of the mineral salts content, have potential for therapeutic spa treatment (REM, 2004).

The Umm er Radhuma Paleocene age carbonate aquifer, widespread throughout the Arabian Peninsula, forms prolific aquifers in Saudi Arabia and Oman, but does not have the same potential in UAE, especially Abu Dhabi Emirate. Very few water wells have penetrated this aquifer in the Emirate and our knowledge of it is restricted mostly to the Southern structural domain, south of Al Ain. Attempts at deep exploration in the western region have been fraught with difficulty and campaigns have been few and largely unsuccessful to date with very high groundwater salinities and low well yields. Only one GWAP well intersected Um er Radhuma limestone; thickness of 200 m between depths of 600–800 m below ground with a TDS content of 185,000 mg/l and a small yield of $12\,m^3/hr$. Further investigation into this aquifer system, however, is warranted.

The Simsima Cretaceous limestone Formation occurs in the Eastern Structural domain region and crops out as outliers near the Al Ain – Dubai highway. Exposures of crystalline, fossiliferous limestone of the Simsima Formation can be found at Jebal Mohayer, Qarn Saba, Jebal Masakin, Qarn Tarab and Qarn Bida bint Saud. Over 40 exploration wells have been drilled into this Formation by GWAP with very variable results dependent on intersection of permeable fractures and joints which provide all the yields from the wells. Where fractures are encountered, Transmissivities of greater than $3{,}000\,m^2/d$ and yields in excess of $5{,}800\,m^3/d$ have been reported.

2.3 *Groundwater quality*

Table 2 provided a summary of the three different kinds of groundwater flow system found in the Emirate. Generally, the longer the residence time, the higher the salinity (Total Dissolved Solids-TDS) content of the indigenous groundwater. TDS is widely taken as an indication of ground water quality and also proximity to either a recharge source or discharge area. TDS is obtained by analysis of waters sampled by various means from generally boreholes or wells. Extreme caution

(GTZ et al, 2005) is required in interpreting the resultant TDS, which is almost always calculated analytically by its relationship with Electrical Conductivity which is measured directly by field instrumentation. There are other factors, besides those mentioned above, which may also have an over-bearing effect on salinity e.g. occurrence of gypsum or anhydrite layers within a succession, which although may be very thin, will have a tremendous effect on increasing the TDS of water sampled from the well.

Various classifications of groundwater type have been used internationally and different classifications are also found associated with different projects and government agencies within Abu Dhabi Emirate. Table 3 shows a summary of the main classifications used in the Emirate, all generally use TDS for classification of groundwater type. The classifications of groundwater quality shown in Table 3 are all more detailed than the World Health Organisation simplistic classification for drinking water which has fresh defined up to 1600 mg/l and brackish above this value. In Abu Dhabi Emirate, the term "Fresh" is generally a local standard for potable water.

The Agriculture Extension Department of Abu Dhabi Municipality & Agriculture classification uses electrical conductivity and classify according to suitability of irrigation of selected crops under pre-selected groundwater salinity classes. Although pollution of groundwater from anthropogenic activities, especially agriculture, does exist within the Emirate, this has not had a significant impact on regional groundwater salinity. Rather, local pollution e.g. Nitrate (ERWDA/FEA, 2005)

Table 3. Summary of groundwater classification schemes used in Abu Dhabi Emirate.

Source	TDS Range (mg/l)	Classification
(ERWDA, 2003)	0–1500	Fresh
	1500–8000	Low Brackish
	8000–15000	High Brackish
	15000–35000	Saline
	>35000	Hypersaline
GTZ et al (2005) – German Standards	0–1500	Fresh
	1500–4000	Slightly Brackish
	4000–7000	Medium Brackish
	7000–10000	Strongly Brackish
	10000–25000	Slightly saline
	25000–50000	Medium Saline
	50000–100000	Strongly Saline
	>100000	Brine
USGS (USGS/NDC,1996)	0–1500	Fresh
	1500–15000	Brackish
	>15000	Saline
Forestry Dept	0–1500	Fresh
Abu Dhabi Municipality & Agriculture	1500–10000	Brackish
	10000–20000	Saline
	>20000	Very Saline
Abu Dhabi Municipality & Agriculture –	0–4000 μS/cm	Class I Fresh
Agriculture Extension Service	4000–8000 μS/cm	Class II low brackish
	8000–12000 μS/cm	Class III high brackish
	>12000 μS/cm	Class IV saline
Al Ain Municipality & Agriculture –	0–1000	Class 1 very fresh
Agriculture Extension Service (2001)	1000–2000	Class 2 fresh
	2000–4000	Class 3 low brackish
	4000–6000	Class 4 medium brackish
	6000–8000	Class 5 high brackish
	>8000	Class 6 saline

of some parts of the surficial aquifers has occurred, and is invariably associated with the use of inorganic fertilizers.

2.4 *Groundwater resources evaluation*

The GWRP (USGS/NDC, 1996) and GWAP (GTZ et al, 2005) have used different methods to calculate the groundwater in storage in the Emirate, but both have ultimately calculated average saturated aquifer thickness and specific yield to estimate stored volumes. The volume of fresh groundwater calculated differs by only 8%. It is not possible to compare the saline and brackish groundwater calculations, since different thresholds have been used to define this water quality. The GWRP calculated a total groundwater reserve of 253 Km^3 (7% fresh, 93% brackish – see Table 4) and the GWAP total estimate of 640 Km^3 (79.4% saline) is much larger since groundwater of salinity of up to 100,000 mg/l TDS was included, whereas the GWRP included groundwater with less than 15,000 mg/l TDS. The most striking feature of both estimates is that the amount of fresh groundwater remaining in storage is very small, ranging from 2.6% to 7% of the total. According to the GWAP assessment more than three-quarters (12.5 km^3) of the fresh water in storage occurs in the Liwa lens and only about 4 Km^3 in the Eastern region. According to the GWRP assessment, at current groundwater abstraction rates, it is projected that the fresh and brackish groundwater resources will be depleted in 50 years.

3 GROUNDWATER DEVELOPMENT

It is estimated that about 2.5 billion cubic meters/year of groundwater is currently abstracted (ERWDA, 2003) and is utilized by the Agriculture (76%), Forestry (23%) and Amenity/Domestic/Industrial sectors (1%).

3.1 *Domestic sector*

Water demand in this sector includes domestic and bulk categories. Domestic water demand includes mainly residential, commercial establishments, hospitals, hotels, offices, and shops. Bulk water demand includes agriculture, landscaping, large industrial usage, palaces, airports, and other non-domestic bulk diversions.

The significant increase in customer demand for water occurred mainly in government sponsored housing development schemes and agricultural activities; particularly in the farming and forestry sectors.

Table 4. Abu Dhabi Emirate groundwater reserves estimate from GWRP (USGS/NDC, 1996).

Salinity zone	Area ($m^2 \times 10^6$)	Average saturated thickness (m)	Average specific yield (%)	Volume stored (km^3)
Fresh – Eastern Region	1,440	20	14	4
Fresh – Western Region	2,400	26	23	14
Fresh – Emirate	3,840	–	–	18
Brackish below Fresh Water – Eastern Region	1,440	40	14	8
Brackish below Fresh Water – Western Region	2,400	69	23	38
Brackish – Remaining areas	29,983	42	15	189
Brackish – Emirate	–	–	–	235
Total Fresh and Brackish Emirate				253 Km^3

In 2003, 15.5% (580 Mm^3) of all water consumed in Abu Dhabi Emirate was in the domestic sector; 96% from desalination, 4% only from groundwater wellfields. All of the 16 producing wellfields, containing 600 wells, of which 333 only are operated, are located in the Eastern Region; in 2002; wellfields in the Liwa area of the Western Region were discontinued for domestic supply due to water quality difficulties. Table 5 shows a classification of the Eastern Region wellfields based on groundwater salinity. Previously, all of Al Ain City's domestic water requirements were met from wellfields, however, massive increases in domestic demands, from an annual population growth rate of up to 8%, has meant that wellfields have been placed under increasing stress, resulting in declining water levels, increase in groundwater salinity with a resultant decrease in total production (Figure 6). The widening gap between groundwater supply and domestic demand has been met from an expansionist policy of desalination using all types of production process under an ever increasing responsibility of the private sector. In 2003, total domestic wellfield production had reduced to only 26 Mm^3/yr, meeting only 17% of the total domestic requirements in the Eastern Region. Since 1998, production from the domestic wellfields has decreased by over 60%. Table 5 shows that a large proportion of abstracted groundwater no longer meets the Abu Dhabi drinking water standard (RSB, 2004) and this challenge has been met by blending indigenous, brackish groundwater with imported desalinated water from the Arabian Gulf and, more recently, from the Gulf of Oman at Qidfa, Al Fujairah.

Despite their costly operation for a relatively small to moderate production, the wellfields continue to operate since they are strategically important as an emergency back up to any potential failure in the supply of imported desalinated water to the Eastern Region. The balance of domestic demand in the Eastern Region and also the full requirements for the Western Region, are now met by desalinated water. The provision of desalinated water supply has been well planned, but extremely costly,

Table 5. Classifications of the Eastern Region domestic supply wellfields (Brook, 2003).

Wellfield name	Water quality classification	% Total 2002 production
Shuwaib South, Al Haiyer, Al Karaa, Ghashaba, Al Zaroub, Khashona, Um Ghafa.	Fresh	64
Shuwaib North, Al Khadar, Bida bint Saud, Al Ashoosh, Jabal Oha, Al Wagan, Al Qua, Seih Al Raheel	Low Brackish	35
Al Aslab	High Brackish	1

Note: Fresh <1000 mg/l TDS, Low Brackish 1000–8000 TDS, High Brackish 8000–15000 TDS.

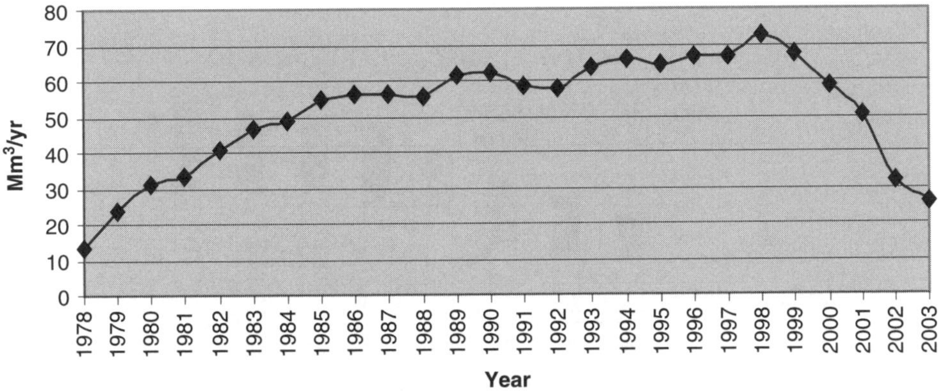

Figure 6. Al Ain wellfields total production 1978–2003.

however supply has always kept abreast of demands, and excess capacity at present provides a potential resource for artificial recharge of aquifers utilizing Aquifer Storage and Recovery (ASR); a relatively new concept for this region, which is currently being tested at the pilot test study level in the Eastern and Western Regions of the Emirate. In 2003, six desalination companies had a combined capacity of 746 Mm^3/yr, 35% of which is currently unutilized.

Using groundwater as a source for the desalination process has been limited to the Al-Ain region. Several reverse osmosis plants have been operated in the Eastern Region, south of Al Ain city at e Um Al Zumol, Al Quaa and Al Wigan. Recently, however, dependence on groundwater for potable water production has been decreased and all plants have been abandoned due to the depletion of the saline water source; except the Um Al Zumol plant, where the production is 36,000 gallons per day.

3.2 *Agricultural sector*

This sector consumes 58% (1949 Mm^3/yr) of all of Abu Dhabi Emirates water demands. By the end of 2003, there were about 25,000 citizen's farms, occupying around 75,500 ha and a small number of large, state fodder (government) farms occupying about 17,000 ha (Brook et al, 2005). Figure 7 shows the expansion in agriculture due to large government support. Citizen's farms are typically 2–3 ha in size and each has two drilled wells at opposite corners of the plot. A well supported system of subsidies promotes agricultural expansion to the tune of 3,000 new farms each year, although expansion is currently restricted due to exhaustion of groundwater supplies. Figure 8 shows a map of all agriculture and forestry developments in Abu Dhabi Emirate. The major limitations on agricultural development are lack of groundwater resources and high groundwater salinity used in irrigation. Close proximity of wells results in well interference effects and unrestricted irrigation causes extreme cones of depression (Figure 8) resulting in a deterioration in salinities, which are usually low brackish to high brackish to begin with.

For example, in citizen's farms in the Al Ain region, irrigation water salinity exceeds 4000 mg/l in 65 percent of farms (ERWDA/MMI, 2004). This can be illustrated by examining the effect of agriculture on the groundwater situation in the following areas, south of Al Ain: In the Al Ouaa/Al Wigan area (Fig 9C) the practice of agriculture and also Forestry in the area has been unsustainable and the groundwater reserves can no longer provide sufficient groundwater of the quantity and quality required for sustaining the agriculture infrastructure that has already been developed. The decline in the groundwater table in this area is shown in Table 6. Investigations in Oman show that the regional through-flow (natural recharge) across the International Border for the Lower Sawamahan Wadi Catchment, and the southern part of Wadi Dank Catchment, both of which incorporate the agriculture areas under study, is of the order of about 4.5 Mm^3/yr. This natural recharge provides a quantity of groundwater which could provide for sustainable farming. However, the rapid expansion of citizens farms (there are now 210 farms alone in Um Al Oush,) over the last 10 years has resulted in massive groundwater abstractions. Citizen's farms alone use around

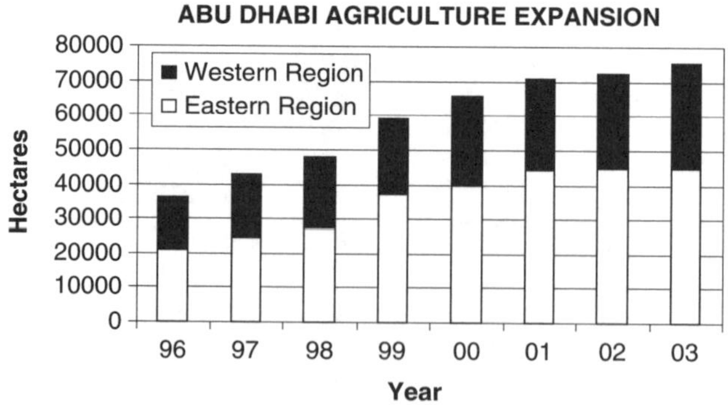

Figure 7. Expansions of citizens farms 1996 to 2003.

66 Mm^3/yr, 15 times that which is naturally replenished to the aquifer by way of through-flow across the International border.

Agriculture is generally dominated by two perennial crops, Dates and Rhodes grass, with some seasonal plantings of short season annual vegetable crops; a limited area of cereals and fruits are also grown. Most agriculture is on small private farms that have been established in relatively

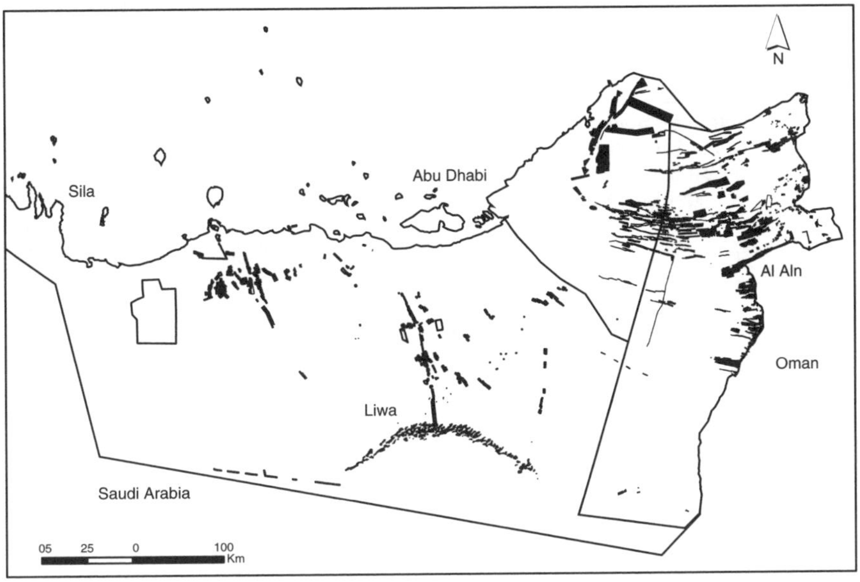

Figure 8. Locations of agriculture and forestry developments.

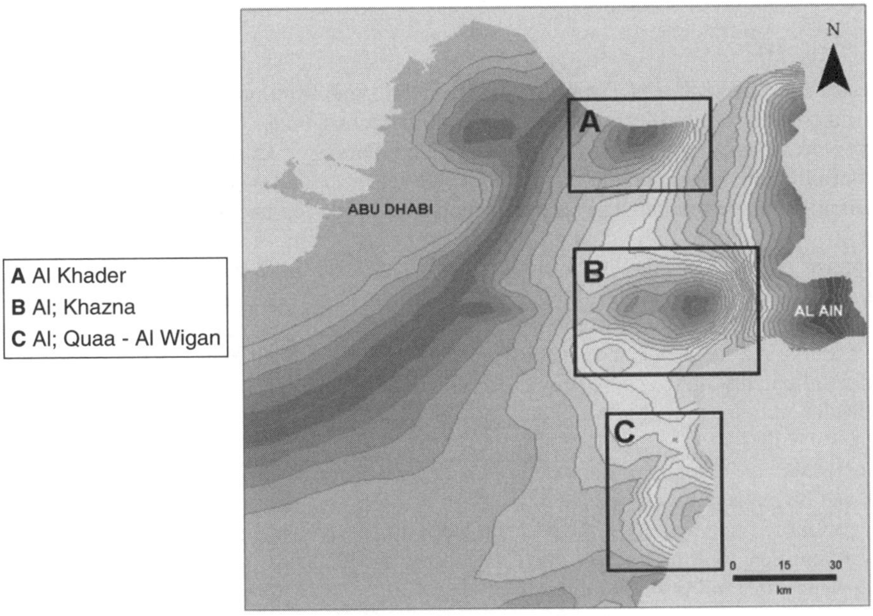

Figure 9. Groundwater cones of depression (January 2005).

Table 6. Changes in groundwater levels over the last 10 years in the Um Al Oush area.

Area	Original static water level (mamsl)	Present static water level (mamsl)	Change in groundwater level (m)
Seah Al Raheil	139.1	50.1	89
Seah Al Regat	126.05	65.11	60.94
Um Al Oush	142.34	49.1	93.24

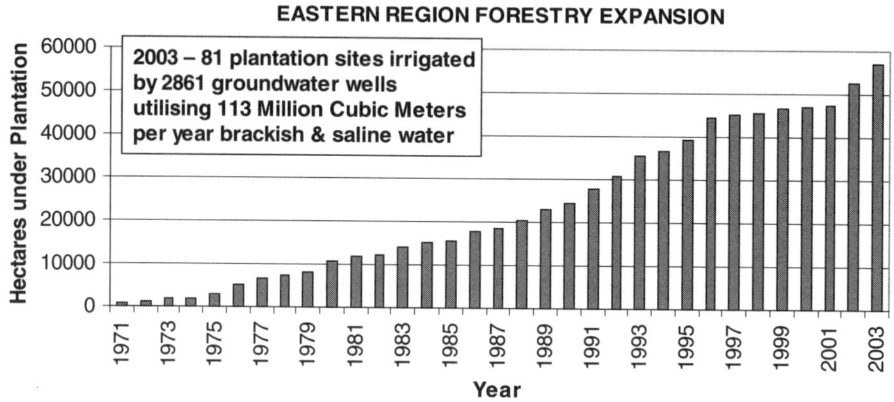

Figure 10. Forestry expansions in the Eastern Region of Abu Dhabi Emirate.

recent times, but there are also small areas of traditional date palm gardens (Al Ain Oasis consume about 10 Mm^3/yr of groundwater for around 375,000 date palm trees and occupy an area of 350 ha) and larger government forage production units. There is also a limited area of protected horticulture (greenhouses, cloches etc).

3.3 *Forestry sector*

This sector consumes 18% (607 Mm^3/yr) of all of Abu Dhabi Emirates water demands. By the end of 2003, there were around 250 separate forestry plantations, under the management of the Al Ain Forestry Department and the Abu Dhabi Municipality Forestry section. Expansion in the sector in the Eastern Region has been rapid and is shown in Figure 10. 58,000 hectares of development was recorded in 1989; the total area under cultivation is now 305243 hectares, representing an expansion of 26% per annum. Individual plantation sizes range from 4 to 70,000 hectares and 80% of development is located in the Western region. A total of 64 million trees are irrigated by 5,713 wells in the Emirate. In the Eastern Region, 12264320 trees, occupying 56854 ha, are irrigated with 122.85 Mm^3/yr. In the Western Region, 51317276 trees, occupying 243494 ha, are irrigated with 484.45 Mm^3/yr. All forestry is irrigated by groundwater; recently, there is a development to supply limited desalinated water to some projects in the western region. This sector is faced with operational challenges related to poor water quality, lack of sufficient quantity of irrigated water and also poor quality soils.

Groundwater used for irrigation ranges in quality from 4200–40,000 mg/l. Acceptable tree growth can be managed with groundwater of 7000 mg/l but since the location of forest development is generally fixed by others, the Forestry Departments sometimes have to make do with water of salinity inimical to good tree growth; an example is Al Yaeela some 55 km WSW of Al Wigan where groundwater salinity was too high (over 35,000 mg/l) for forest establishment to proceed.

In the Western Region, salinities of mostly over 10,000 mg/l and up to 40,000 mg/l are found; this is supplemented by fresh groundwater imported from Khashona wellfield in the Eastern

Region and recently, desalinated water from Taweelah. Even with blending from these lower salinity sources, only rarely is water with a quality as good as 5000 mg/l used. Over 17,000 ha of forest developments exist along the Abu Dhabi– Dubai road. A 1000 ha forest in the Al Samha area uses desalinated water and in Al Shehamah forest, a mix of desalinated water and groundwater is used. Both groundwater quality and well yields are reported to be slowly deteriorating and increased use of desalinated water is foreseen.

In order to optimize tree growth potential under these difficult conditions, emphasis has been placed on growing indigenous species such as *Prosopis cineraria (ghaf) and, Salvadora persica*), etc. In the Eastern Region, these two species alone account for over 60% of all trees.

3.4 *Amenity plantation sector*

Amenity irrigation for parks, gardens and recreational areas e.g. golf courses, football pitches etc. accounts for just over 7% of total consumption. This sector relies mostly on treated effluent as a source, but wells are also utilized, especially in Al Ain City, where over 400 have recently been inventoried. Combined, it is estimated that amenity plantations use about 245 Mm^3/yr of water, a combination of treated effluent, desalinated water and also local groundwater. Artificial lakes and ponds have been created throughout the Emirate using groundwater to enhance the visual appearance of the desert and to improve the recreational appeal of parks and gardens. Two of note is seen at Ain Al Fayda and Al Mubazzarah Lake, both in the Al Ain area. At the former, the spring channels and ponds have been dredged for recreational boating purposes. At the latter, a lined lake has been created to collect outflows from pumped wells within the Jebel Hafit area which provide geothermal waters which are delivered through channels to open – air Jacuzzis for public use. The lake also provides for boating activities and at both venues there are open air swimming pools supplied by natural groundwaters. The thermal, mineral groundwaters at Jebel Hafit are thought to have potential therapeutic value (REM, 2004). Mubazzarah waters have been compared with 450 other mineral water samples and four existing thermal centers or spas, in Tunisia, Morocco and France, all have similar chemical analyses.

3.5 *Industrial/commercial sector*

Detailed information on the source and use of water for this sector is not readily available and is poorly documented. The sector is currently estimated to use only 1.5% of the total water consumed in the Emirate, but the proportion will increase as expansion in the Industrial sector is brought about by the development of a number of new Industrial Estates in Abu Dhabi, Al Ain and elsewhere. In the Al Ain area, only two industries have been identified which have an independent water supply from pumped boreholes linked where appropriate, to desalination units on site; these are the Al Ain Mineral Water Company and the Coca Cola bottling plant. Apart from these exceptional arrangements, it follows that all industrial demand is being met from bulk water supplied by the distribution companies.

4 MAJOR WATER RESOURCES ISSUES & CHALLENGES

The following major water management issues have been identified as part of ongoing EAD studies in developing a water resources management strategy and action plan for the Emirate of Abu Dhabi:

Water use: policy, planning and regulation

- Reduction in quantity and quality of groundwater through over-abstraction, resulting in salinization of land, reduction in crop yields and abandonment of farms
- Lack of farm management leading to over- irrigation and drainage problems

- Unplanned development in the farming & forestry sectors
- Poor performance of the Forestry Sector due to insufficient water and poor water quality
- Little or no effort to manage the demands for water in agriculture sector
- Lack of recognition of the true economic cost of water when assigning its use
- Uncontrolled and un-regulated well drilling, leading to dry wells and wasted resources

Protection, conservation and monitoring of water resources

- Lack of a coordinated Emirate-wide water resources monitoring network and programme
- Groundwater pollution due to fertilizer use
- Lack of groundwater protection policies, e.g. no protection zones for municipal wellfields that still produce water of drinking quality, for example
- Lack of inventories on sources and demands
- Lack of qualified, technical, on-site supervision for drilling water wells
- Poor monitoring and data collection during drilling
- Insufficient water resources monitoring
- General waste of water and leakages
- Poor practices of water data and information management
- Non-availability or poor access to water resources information and data, and lack of a central, Emirate-wide database to hold and analyse water resources data and information
- No well inventory, poor data collection when drilling wells

Coordination of groundwater exploration and assessment

- Need for expansion of groundwater exploration programmes, especially for deeper aquifer potential
- Lack of coordination and collaboration between existing groundwater exploration and assessment programmes

Local, regional & International cooperation and collaboration

- Little or no technical cooperation with neighbouring Emirates and countries, especially on developments on or near to the international boundaries

Strategic emergency water resources

- No developed strategic reserve of potable quality water in case of emergency (current reserve for less than 2 days)

Common to the solution of most of the issues and problems listed above is the requirement for the establishment of a central, independent authority for Water Management in Abu Dhabi Emirate. Up until recently, responsibility for managing water sources and water use was divided between several organizations and agencies. This fragmented arrangement is unsatisfactory for effective water management and results in lack of coordination and collaboration between some of the bodies, duplication of efforts, non-assignment of responsibility for some very important aspects of water management, wasted funds and a general lack of accountability for some organizations current practices which go against the principle of sustainable water resources management and rationalization of water use.

5 OUTLOOK

The outlook for water resources management generally within the Emirate of Abu Dhabi is bright; this positive statement is made mostly because the responsibilities for managing the various aspects of water supply and use are now clearly defined amongst the various Abu Dhabi Government agencies and there is now a real opportunity to implement effective programmes as part of an overall strategy for water resources management and development within the Emirate. This clear identification of roles and responsibilities has not always been the case, with the effect of duplication of effort, ineffective control and management and also wasted resources. The list of current responsibilities for the various Government Agencies is shown in Table 7. In March, 2005, the Environment Agency Abu Dhabi (EAD) was given responsibility for the management of groundwater. EAD's main focus at the moment is on the following areas:

Regulation of groundwater use

- Inventory of existing wells
- Registration of existing and new wells
- Detailed well drilling and abstraction permitting policy
- Establishment of a Law on water well drilling (Draft already completed)
- Registration of Water well drilling and associated water works Contractors
- Registration of Water well drilling and associated water works Consultants.

Groundwater resources monitoring system

At present, no national groundwater monitoring system exists for Abu Dhabi Emirate (Brook, 2005), and has never existed. In March 2005, the GTZ Groundwater Assessment Project and associated groundwater monitoring programme was transferred to EAD. This very comprehensive network of observation boreholes, and those currently operated by the GWRP forms a very good basis for a National Groundwater Monitoring Network (Figure 11).

Groundwater information network

Establishing a centralized water resources database, using the existing (Dawoud et al, 2005) as a base for the following aspects: (1) exploration; (2) monitoring; (3) water use; (4) water supply;

Table 7. Current responsibilities in the water sector.

Ser No.	Government agency	Responsibility
1	ADWEA	Supply and Distribution of Drinking Water
2	EAD	Management, monitoring, assessment and regulation of Groundwater and Protection against pollution
3	Municipalities & Agriculture	Development of Agriculture Irrigation
4	Municipalities & Agriculture (Abu Dhabi)	Development of Forestry Irrigation
5	Dewan of Eastern Region	Development of Forestry Irrigation
6	Municipalities & Agriculture (Al Ain)	Management of Sewerage and waste water treatment Eastern Region
7	Municipalities & Agriculture (Abu Dhabi) Sewage Projects Committee	Management of Sewerage and waste water treatment Western Region
8	Regulation and Supervision Bureau	Regulation of drinking water and sewerage/Waste water treatment
9	Ministry of Communications	Meteorological monitoring and assessment
10	Ministry of Presidential Affairs Dept of Atmospheric Studies (formerly DWRS)	Meteorological monitoring and assessment
11	ADNOC/NDC/USGS	Specialist Groundwater Research (Eastern Region)

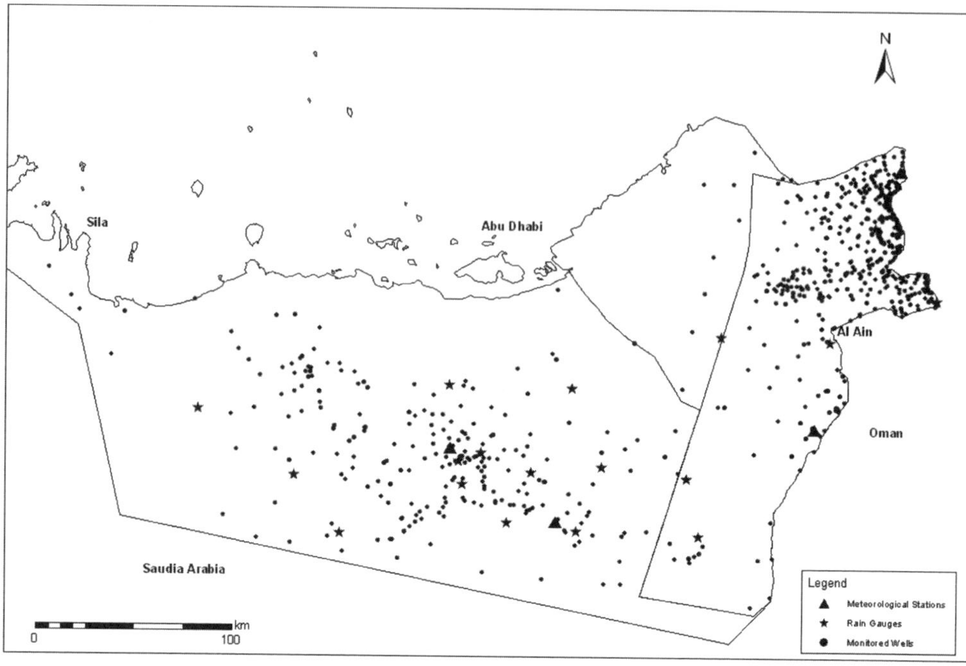

Figure 11. Current monitoring well network.

(5) well permitting; (6) well registration; (7) contractor & consultant registration; (8) assessment; (9) modelling; (10) meteorology; and (11) mapping.

The water information obtained will eventually be accessible on – line as part of EAD's current commitment to achieving e-service government status by the end of 2006.

Strategic Water Reserves

A committee for Artificial Storage and Recovery Schemes in the Emirate has been formed and EAD now has the overall responsibility for managing pilot ASR projects in the Eastern and Western Regions which will both be completed by the end of 2006.

ACKNOWLEDGMENTS

The author is grateful to H.H. Sheikh Khalifa bin Zayed Al Nahyan, President of the UAE and Chairman of the Governing Board of the Environmental Agency Abu Dhabi (EAD) and to H.H. Sheikh Hamdan bin Zayed Al Nahyan, Deputy Chairman of the Governing Board, for financing this sector paper. I also wish to thank HE Mohammed Al Bowardi, Managing Director of EAD, Mr. Majid Al Mansouri, Secretary General, ERWDA and Mr.Abdulnasser Al Shamsi, Director Terrestrial Environment Research Center (TERC) for their invaluable support.

REFERENCES

Akbar, M., 1994, Structural analysis of borehole electrical images on wells GWP-134, 135, 136, & 137: Consultant's report by Schlumberger-CGD, Abu Dhabi, January 1994, 29 p.

Akbar, M., Bright, D.J., Mohamed, F.A., and Nurmi, R.D., 1995, Anticlinal reservoir characterization and modeling using outcrop, seismic, and borehole data: Unpublished *journal article* on file with National Drilling Company, [16] p.

Al Ain Municipality & Agriculture – Agriculture Extension Service 2001, Annual Report on Groundwater Salinity.

Bright, D.J., and Silva, Eric, 1998, Hydrogeology of the surficial aquifer system in northeastern Abu Dhabi, United Arab Emirates: U.S. Geological Survey Administrative Report, 2 sheets, prepared for the National Drilling Company, Abu Dhabi.

Brook, M.C.,1994, Preliminary Assessment of the Fars Aquifer in the Dhahirah Region of Oman. Ministry of Water Resources Technical Report.

Brook, M.C., 2001, History of the Al Masarrat Aquifer Water Supply Scheme, Sultanate of Oman (Proc. Geology of Oman Conference, Oman 2001).

Brook, M.C 2003, History & Current Status of Municipal Wellfields in the Eastern Region of Abu Dhabi Emirate, 1979–2002. Vol I & 11.

Brook, M.C, Al Houqani, H., Dawoud, M. 2005, The Opportunities for Water Resources Management in the Emirate of Abu Dhabi, United Arab Emirates. In Proceedings of 7th Gulf Water Conference, Kuwait, 2005.

Brook, M.C, 2005, Development of a Groundwater Monitoring Network for Abu Dhabi Emirate.

Czarnecki, J.B., Sanford, W. and Wood, W.W. 2000, Brine Resources in Coastal Sabkhas of Abu Dhabi, United Arab Emirates: U.S. Geological Report Administrative Report, July 2000, 98p., Prepared for Abu Dhabi National Oil Company (ADNOC).

Dawoud, M.A., Farbridge, R., Wonderen, J. van, Brook, M.C. 7 Al Houqani, H. 2005, An ARCGIS database for water supply/demand modeling and management in Abu Dhabi Emirate, UAE. *WASTA Seven Gulf Water Conference Proceedings*: Water in the GCC-Towards integrated Management: 575–592, Kuwait.

ERWDA, 2003, Water Resources Statistics bulletin for Abu Dhabi Emirate 2003.

ERWDA/Mott Macdonald International (MMI), 2004, Preliminary Assessment of the Water Situation in the Eastern and Central Regions of Abu Dhabi Emirate.

ERWDA/FEA 2005, Assessment of residual Pesticide and Fertiliser Residues in groundwater in Abu Dhabi Emirate. Final Report.

Fitterman, D.V., Menges, C.M., Al Kamali, A.M., and Jama, F.E., 1991, Electromagnetic mapping of buried paleochannels in eastern Abu Dhabi Emirate, U.A.E.: *Geoexploration*, V. 27, p. 111–133, Elsevier Science Publishers B.V., Amsterdam.

GTZ/Dornier Consult/ADNOC 2005, Status Report Phases 1Xa, 1Xb and 1Xc for Groundwater Assessment Project Abu Dhabi.

Imes, J.L., Hutchinson, C.B., Signor, D.C., Tamayo, J.M., Mohamed, F.A., and Hadley, D.G., 1994, Ground-water resources of the Liwa Crescent area, Abu Dhabi Emirate: U.S. Geological Survey Administrative Report 94–001, 138 p., prepared for the National Drilling Company, Abu Dhabi.

Khalifa, M.A., 1997, Hydrogeology of the geothermal fractured-rock well field at Jabal Hafit, Abu Dhabi Emirate: Water Sciences & Technology Association, *Proceedings of the Third Gulf Water Conference*, March 8–13, Muscat, Oman, V. 1, Water Resources Planning and Management, p. 125–140.

Khalifa, M.A., 2004, Hydrogeologic setting and characterization of the aquifer system in Al Wigan area, South Al Ain, UAE. UAE University, Deanship of Graduate Studies, Master of Water Resources.

Kirkham, A, 2004, The Geology of Jebel Hafit in Jebel Hafit: A Natural History. Edited by Aspinall. S and Hellyer, P.

Moreland, J.A., 1998, Activities, findings, and conclusions of the Ground-Water Research Project for the Emirate of Abu Dhabi 1988–1997: U.S. Geological Survey Administrative Report, 34 p., prepared for the National Drilling Company, Abu Dhabi.

National Drilling Company (NDC) – United States Geological Survey (USGS) 1993, Groundwater Research Project. Groundwater Resources of Al Ain area, Abu Dhabi Emirate. US Geological Survey administrative report prepared in cooperation with NDC, October 1993.(Phase II).

Resources en Eaux Minerales (REM) 2004, Qualitative valuation of mineral water resources in the area of Al Ain, UAE.

Rizk, Z.S, Garamoon, H.K, El Etr H.A. 1998, Contribution to hydro-geochemistry of the Quaternary aquifer in the Al Ain area, UAE. Proceedings of the International Conference on Quaternary deserts and climate change, Al Ain, UAE. In Quaternary deserts and climate change. Alshrhan, A.S, Glennie, K.W., Whittle, G.L & Kendall, C.G. ST. C.A.A Balkema, Rotterdam.

Rizk, Z.S., Alsharhan, 2003, Hydrogeology, groundwater chemistry and isotope hydrology of the Quaternary Liwa aquifer in the western region of the United Arab Emirates. In Proceedings of sixth Gulf Water Conference, Riyadh, Saudi Arabia.

RSB 2004, Regulation and Supervision Bureau for the Water and Electricity Sector in the Emirate of Abu Dhabi. Water Quality Regulations, Document ED/R01/001 (revision 2).

Symonds, R., Robledo, A and Al Shateri, H, 2005, Use of environmental tracers to identify and date recent recharge to the surficial aquifer of northeastern Abu Dhabi Emirate, United Arab Emirates: NDC – USGS Technical Series Administrative Report 2005–001, 24p.

UAE University 1993, Atlas for United Arab Emirates.

US Geological Survey/NDC Administrative Report 1994. Groundwater Resources of the Liwa Crescent Area, Abu Dhabi Emirate.

US Geological Survey (USGS) 1996, Groundwater resources of Abu Dhabi Emirate.

Wood, W.W., and Imes, J.L., 1995, How wet is wet? Precipitation constraints on late Quaternary climate in the southern Arabian Peninsula: *Journal of Hydrology*, V. 164, p. 263–268.

Wood. W.W. & Sanford, W.E., 2002, Hydrogeology and solute chemistry of the coastal – sabkha aquifer in the Emirate of Abu Dhabi. In Sabkha Ecosystem, Vol 1- The Sabkhas of the Arabian Peninsula and Adjacent countries. Hans – Joerg Barth & Benno Boer, Eds, Kluwer Academic Publishers, Dordrecht, 354p.

Woodward, D.G., 1990, Use of uphole velocity surveys for mapping hydrogeologic boundaries, Abu Dhabi, United Arab Emirates [abs.]: Abstracts for the AGU Fall Meeting, Supplement to *Eos, Transactions, American Geophysical Union*, V. 71, no. 43, p. 1,300.

Woodward, D.G., and Al-Jeelani, A.H., 1993, Application of reprocessed seismic sections from petroleum exploration surveys for groundwater studies, eastern Abu Dhabi, UAE: *Proceedings of the 8th Middle East Oil Show & Conference*, SPE 25538 Society of Petroleum Engineers Annual Meeting, April 3–6, Manama, Bahrain, V. 1, p. 257–266.

Woodward, Dennis, and Menges, C.M., 1991, Delineating aquifers in buried paleochannels using uphole velocity profiles from petroleum seismic surveys, northeastern Abu Dhabi, United Arab Emirates [abs.]: Geological Society of America, Abstracts with Programs, 1991 Annual Meeting, V. 23, no. 5, p. A123.

Arid Land Hydrogeology: In Search of a Solution to a Threatened Resource – Mohamed(ed.)
© 2006 Taylor & Francis Group, London, ISBN 0 415 41127 0

Artificial recharge of groundwater: Field experiment

Mohsen M. Sherif
Civil and Environmental Engineering Dept., College of Engineering, UAE University, Al Ain, UAE

Anvar Kacimove
Department of Soils and Water Sciences, College of Agriculture, Sultan Qaboos University, Muscat, Sultanate of Oman

Salim F. Akram & Ampar V. Shetty
Ministry of Agriculture and Fisheries, Dubai, UAE

ABSTRACT: This paper presents the initial results of a recharge experiment that has been conducted in Wadi Ham, UAE. The system is composed of a 25 × 25 m open recharge basin and 8 monitoring wells located in the vicinity of the basin. To provide more control on the recharge conditions, the basin was then subdivided into a number of sub-basins. The initial results indicated very high infiltration rates with a low mound height. The site was found to be suitable for implementation of large scale recharge projects.

1 INTRODUCTION

In arid and semi-arid regions, surface water resources are mostly absent and groundwater resources are often nonrenewable. The scarcity of fresh water resources in such areas is becoming more pronounced. Groundwater levels are declining continuously due to the tremendous increase in the pumping of groundwater to meet the increasing water demands over the last few decades. On the other hand, the quality of available groundwater resources has also deteriorated. Many pumping fields have, therefore, been terminated.

Artificial recharge of groundwater provides a feasible solution to sustain the groundwater systems in arid and semi-arid regions. It has been implemented in many countries around the world. However, because of the heterogeneous nature of groundwater systems that may significantly vary from one location to another, field experiments are always essential to examine the applicability and feasibility of such artificial recharge systems.

Groundwater modeling can be defined as the process of simulating the response and behavior of groundwater systems under natural or artificial recharge/discharge events. Numerical models can be used to provide insight into the controlling parameters and system dynamics in a site specific setting. Models can also be used to study the processes of interaction between different water bodies such as surface water reservoirs and groundwater systems. In this report, the MODFLOW was used to simulate impact of recharge due to a small pond of 20 m by 20.

Upon the progress in the simulation process, MODFLOW was found more convenient to handle the required simulations that require frequent changes in the discharge and recharge events through the simulation time. MODFLOW is a three-dimensional finite-difference ground-water flow model. It has a modular structure that allows it to be easily modified to adapt the code for a particular application. MODFLOW is currently the most used numerical model in the U.S. Geological Survey for groundwater flow problems.

MODFLOW is designed to simulate groundwater systems in which saturated-flow conditions exist, Darcy's law applies, the density of ground water is constant, and the principal directions of horizontal hydraulic conductivity or transmissivity do not vary within the system. However, Hydraulic conductivities or transmissivities for any layer may differ spatially and be anisotropic (restricted to having the principal direction aligned with the grid axes and the anisotropy ratio between horizontal coordinate directions is fixed in any one layer), and the storage coefficient may be heterogeneous. These conditions are met for many aquifer systems for which there is an interest in analysis of ground-water flow and contaminant movement. For these systems, MODFLOW can simulate a wide variety of hydrologic features and processes. Steady-state and transient flow can be simulated in unconfined aquifers, confined aquifers, and confining units. A variety of features and processes such as rivers, streams, drains, springs, reservoirs, wells, evapotranspiration, and recharge from precipitation and irrigation also can be simulated.

2 MODEL GEOMETRY AND GRID

The model domain of Wadi Ham aquifer comprises an area of 1.0 km^2 with total length of 1.0 km east to west (Dam to coast of Oman Gulf) and 1.0 km north to south (Fujairah to Khalba) as shown in the Figure 1. The model area and the aquifer boundaries were delineated and fix the UTM coordinates by digitizing remote sensing image of Wadi Ham area. At the coast side cells which are located close to the BHF-17 and the western side cells are considered to be constant head cells. Pond area, observation wells and pumping wells were located and marked on the study domain.

The study area was divided into 114 columns and 114 rows. The size of the cell is 10 m by 10 m for the model area away from the pond and 5 m by 5 m for the area in and around the pond. A stress period is defined as a time period during which all time dependent processes such as pumping and recharge are constant. The recharge experiment was conducted during the period October 4, 2004

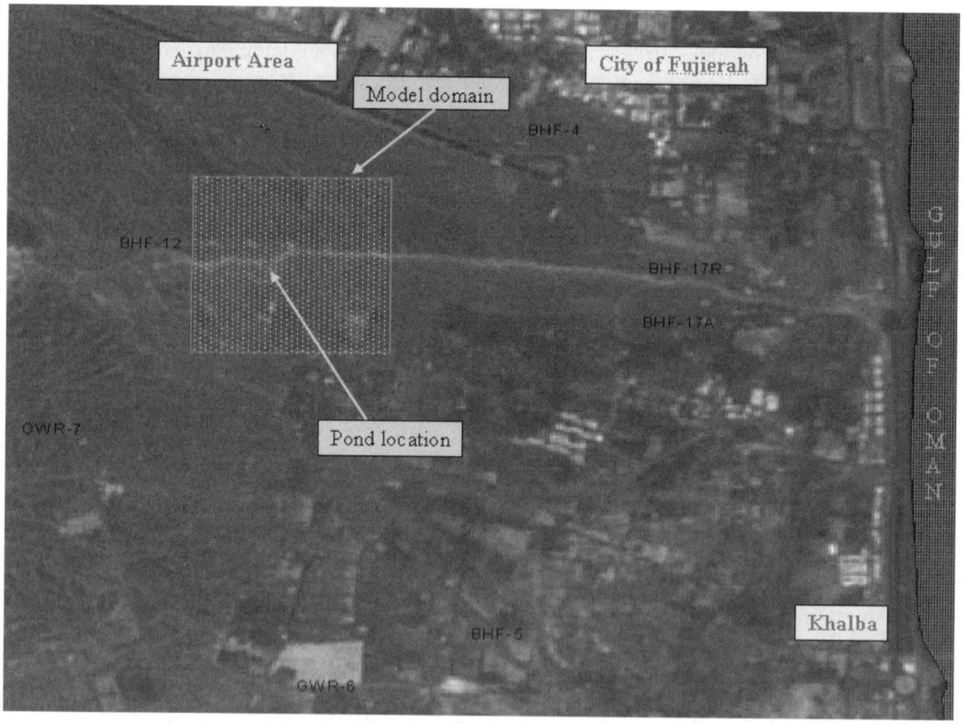

Figure 1. Location of the recharge pond in Wadi Ham.

to November 21, 2004 for 48 days. Other experiments were also conducted from January 4, 2005 to January 13, 2005 for 10 days and from 9 and 25 May 2005 for a duration of 16 days. The results presented in this paper are limited to the first experiment.

2.1 *Input data*

Data regarding pond infiltration, rainfall, observation wells (8), aquifer parameters, and well field and pumping were assembled and prepared in the model input format.

Recharge: The recharge to the aquifer was assigned through two zones corresponding to the recharge from rainfall and from the pond storage/infiltration for the modeling period.
Rainfall: Fujairah rainfall data during the first experiment was considered as it is very close to the experiment site. The rainfall m/day will be assigned to all the active cells on daily basis in the model area.
Pond storage: For the numerical modeling, the storage depth was distributed in space to the 16 cells within the pond area (400 m^2) of pond and in time to the total period of storage as m/day.
Evapotranspiration: The evapotranspiration package of MODFLOW simulates the effect of plant transpiration, direct evaporation, and seepage at the ground surface by removing water from the saturated groundwater regime. Evapotranspiration package requires evapotranspiration rate and extinction depth. The evapotranspiration loss from the water table occurs at the maximum rate when the water table at or above ground surface. The evapotranspiration loss from the water table is negligible when the elevation of water table is below extinction depth. The evapotranspiration loss from the water table varies linearly with the water table elevation between ground surface and extinction depth.

2.2 *Boundary conditions*

Constant head boundary condition is used to fix the head value in selected grid cell regardless of the system conditions in the surrounding grid cells. These cells are acting as an infinite source of water entering the system, or as an infinite sink for water leaving the system. The area occupied by the western and eastern side of the study domain was considered as a constant head boundary cells in the model with head well BHF-12 and BHF-17R respectively for modeling period.

2.3 *Observation wells*

The MODFLOW contains an enhancement design to make model selection more efficient by the comparison of simulated heads with observed heads for every time step. Daily groundwater level data for 8 observation wells around the as shown in the Figure 2 were available for the modeling period.

3 AQUIFER PARAMETERS

Based on the geological investigations, it is found that the system is mainly a quaternary aquifer which is composed of wadi gravels. The gravels are highly permeable with variable hydraulic conductivities. Aquifer parameters have been readjusted during the model run.

3.1 *Aquifer layers*

The model area of lower plains of the wadi Ham composed of recent Pleistocene wadi gravels. This layer is underlain by the consolidated rocks of the Semail formation (Ophiolitic sequence). The thickness of wadi gravel varies from 74 m to 77 m in and around the pond. Figure 3 shows the geological cross-section along the Wadi Ham which includes the model area.

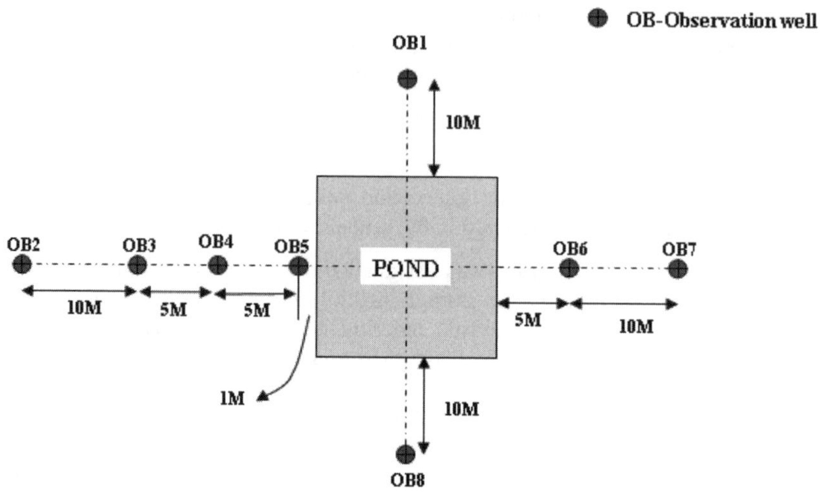

Figure 2. Location of the observation wells in the vicinity of the recharge pond.

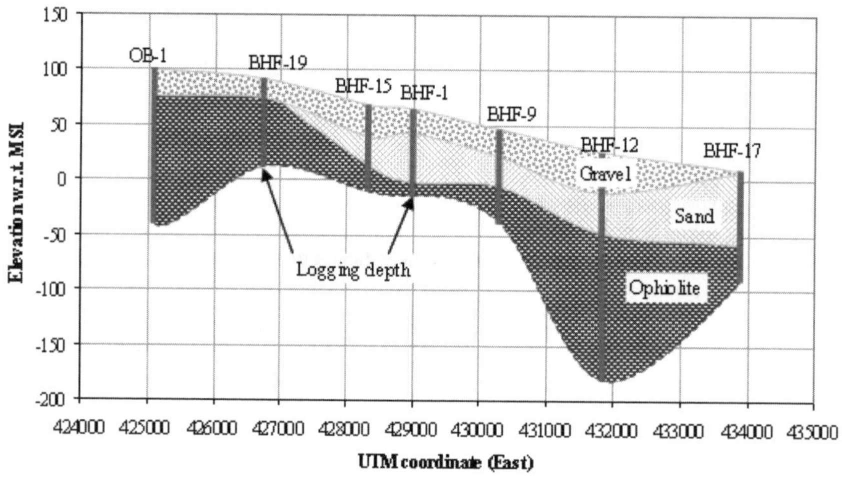

Figure 3. Typical geological cross section along the course of Wadi Ham.

Table 1. Data related to the ground surface elevation and bottom of aquifer layer.

Location	UTM coordinate		Ground surface elevation, msl	Bottom of the layer, msl
	Easting	Northing		
BHF-12	431865	2773432	27.00	−50
Western point	431800	2776300	29	−45
Eastern point	432700	2776300	20	−56
Pond location	432169	2776312	22	−53

Based on the available geological information, the top layer is gravel and sand and the lower layer is of ophiolite. The bottom layer of ophiolite is impermeable in nature and considered to be a non-aquifer. Therefore one layer model of wadi gravel and sand is considered. The data points for the ground surface elevation and bottom of the layer are presented in Table 1.

3.2 *Groundwater abstraction*

Groundwater is exploited intensively from the sand and gravel aquifer for irrigation in the coastal plain close to the pond (Kalbha well field). The well field with about 60 wells is operated since 1995 near Kalbha. The total draft is about 6 million m^3/year.

3.3 *Initial heads*

MODFLOW requires an initial guess for the head values in the model. A proper initial guess for the starting heads of the simulation can reduce the required run time significantly. The initial head values are also used to calculate the drawdown values. A stable piezometric surface head may be obtained by the steady state simulation or available ground water level of the study domain.

4 SIMULATION OF THE RECHARGE EXPERIMENT

The experiment of pond recharge and its simulation was selected from October 4, 2004 to November 21, 2004 for 48 days. A stress period is defined as a time period during which all time dependent processes such as pumping and recharge are constant and could not be changed by the user. The length of stress period in this exercise was taken as one day. The total number of stress periods considered for the calibration period was 48.

The finite-difference technique employed by the MODFLOW requires each stress period to be discretized in to several time steps to obtain an accurate solution. The smaller the time steps, the more accurate the solution obtained. The time step multiplier is a factor that can be used to increment the time step size within each stress period. A time step multiplier value greater than one will produce smaller time steps at the beginning of a stress period resulting in better representation of the changes of the transient flow field. Thus increase in the number of time steps in a simulation may result in smoother head or drawdown versus time curves. Therefore 10 time steps in each stress period with a time step multiplier of 1.2 were considered in the calibration period.

4.1 *Rainfall*

The rainfall of Fuierah Airport was considered for the period. The rainfall depth 10 mm on October 20, 2005 during the experiment period was introduced in the model. The rainfall m/day was assigned to all active cells in the model area.

4.2 *Pond storage*

The continuous discharge ranging from 12 m^3/hr to16 m^3/hr was maintained to the pond during experiment. However, it was observed that no storage created in the pond. Therefore under such circumstances, it is decided to use recharge well instead of storage due to ponding. Recharge rate considered for the experimental period is presented in the Table 2.

4.3 *Evapotranspiration*

Evapotranspiration of 0.014 m/day and extinction depth of 2 m were considered for the model domain.

4.4 *Boundary conditions*

The constant head of western side of the model domain derived from the existing head (−1.78, −1.8) of well BHF-12, where as eastern side model boundary was estimated from the available heads (0.125, 0.15) at pond and BHF-17R during the simulation period.

Table 2. Pond recharge during the first experiment.

Start period	End period	Recharge m^3/hr	Recharge m^3/day
0	4	16	384
4	6	0	0
6	20	12	288
20	48	0	0

Table 3. Observation well record during the experiment.

Stress period	OB-1	OB-2	OB-3	OB-4	OB-5	OB-6	OB-7	OB-8
0	−2.342	−2.63	−2.6	−2.61	−2.63	−2.49	−2.5	−2.7
1	−2.362	−2.63	−2.62	−2.61	−2.66	−2.51	−2.52	−2.72
3	−2.192	−2.59	−2.56	−2.55	−2.2	−2.42	−2.48	−2.65
5	−2.1	−2.53	−2.47	−2.38	−2.48	−2.32	−2.33	−2.57
7	−2.192	−2.44	−2.5	−2.57	−2.52	−2.35	−2.4	−2.59
9	−2.242	−2.51	−2.56	−2.62	−2.63	−2.39	−2.43	−2.64
10	−2.23	−2.58	−2.53	−2.54	−2.55	−2.39	−2.42	−2.64
13	−2.232	−2.57	−2.53	−2.53	−2.52	−2.37	−2.41	−2.62
15	−2.232	−2.56	−2.52	−2.53	−2.53	−2.37	−2.41	−2.62
17	−2.172	−2.47	−2.43	−2.44	−2.45	−2.29	−2.33	−2.53
19	−2.212	−2.54	−2.5	−2.51	−2.51	−2.35	−2.39	−2.6
21	−2.202	−2.58	−2.55	−2.56	−2.57	−2.42	−2.44	−2.65
26	−2.372	−2.67	−2.65	−2.66	−2.69	−2.52	−2.54	−2.74
28	−2.422	−2.67	−2.65	−2.66	−2.69	−2.52	−2.55	−2.75
48	−2.362	−2.64	−2.62	−2.63	−2.67	−2.53	−2.54	−2.73

Table 4. Aquifer parameters adopted for the simulation.

Specific yield	Permeability (m/day) K_x	K_y	K_z
0.1900623	8.2	8.25	0.864

4.5 *Initial conditions*

The most common type of transient flow calibration begins the simulation from the calibrated steady state solution to derive a stable initial head conditions. Steady state condition was performed just before the experiment as there was a long spell of dry condition existed.

4.6 *Observation well data*

The data of 8 observation wells were available around the pond. The detailed water table record for the corresponding stress period is presented in the Table 3.

5 SIMULATED PARAMETERS AND RESULTS

The pond recharge simulation was achieved by changing three parameters, namely, permeability, specific yield and pumping rates. The final values of the specific yield and permeability are presented in the Table 4.

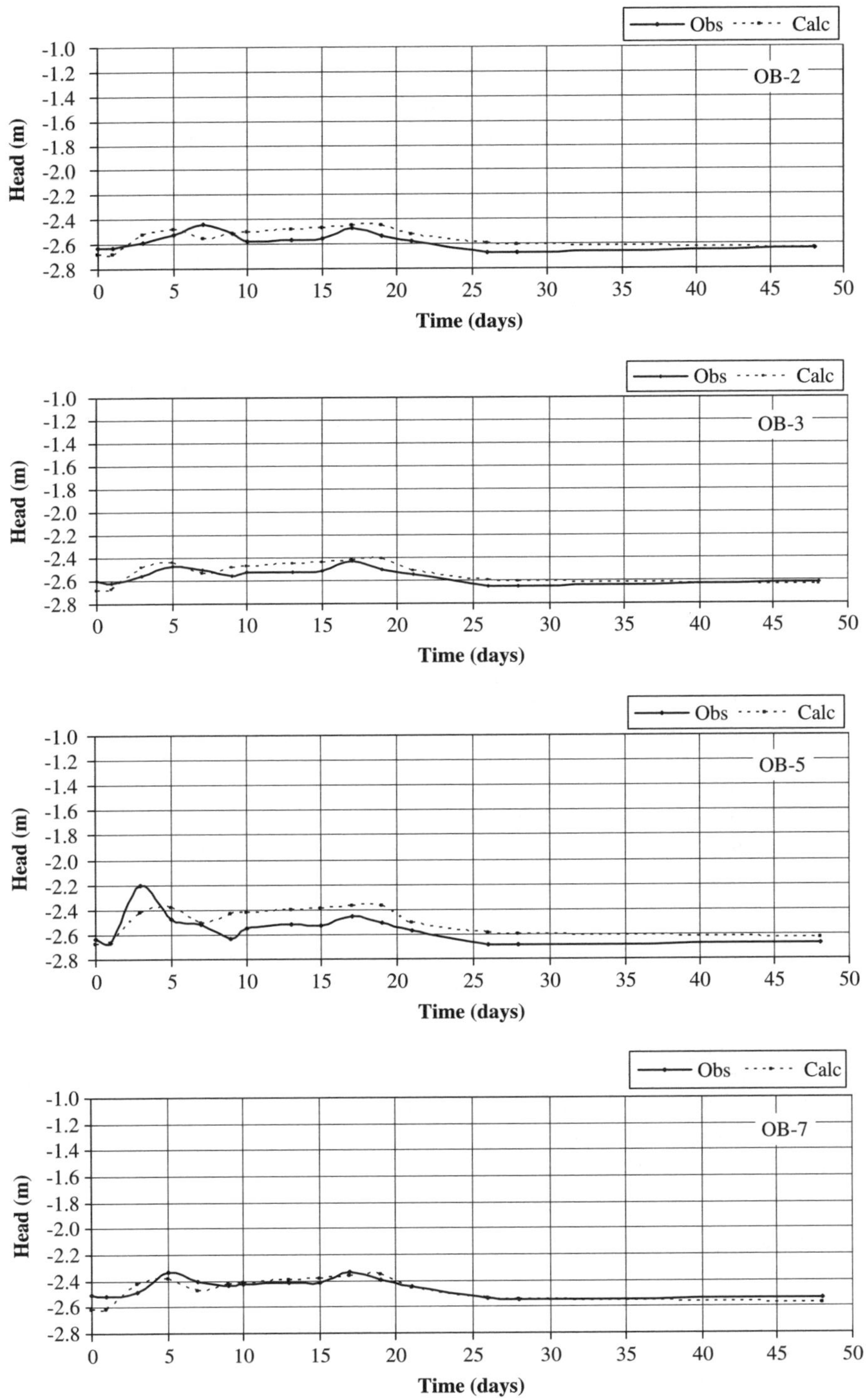

Figure 4. Comparison between simulated and observed water levels in monitoring wells.

Abstraction within the study domain was also simulated by a number of computer runs till the desirable calculated head in each observation wells were achieved. The pumping rate that has been adopted during simulation period is 8750 m^3/day.

The recharge due to rainfall events in the vicinity of the Wadi Ham area was taken about 20 per cent of the total rainfall. This rainfall recharge was calibrated so that the calculated heads at observation points using MODFLOW reasonably match the field measurements.

The numerical simulation was performed under the unsteady conditions to study the groundwater levels and water mound in the vicinity of the recharge basin in space and time. Time series graphs of simulated and observed groundwater levels are shown in Figure 4. As illustrated by these Figures, the resulted groundwater levels are good agreement with the observed levels in the monitoring wells. However, it should be noted that due to the uncontrolled pumping from the area of Wadi Ham some discrepancies were observed.

The initial assessment of the results indicated that the infiltration rate in the selected area is very high and is in the order of 8–12 m/d. On the other hand, the groundwater mound was relatively small, in the order of 40 cm. Therefore, it is concluded that hydraulic conductivities in three directions, K_x, K_y and K_z, are relatively high. Further recharge experiments are needed to provide a quantitative assessment of the long-term infiltration rate.

6 CONCLUSIONS

A field experiment of groundwater recharge was conducted in the area of Wadi Ham to assess the feasibility of the site and identify the infiltration rate. The recharge scheme included a 25 ×25 m recharge pond and 8 monitoring wells surrounding the pond. The recharge water was supplied from the Kalbah well-field which is located about 800 m from the pond. MODFLOW was used to simulate the recharge process. A good agreement was achieved between the simulated and observed groundwater levels.

It is concluded that the selected site is suitable for the application of artificial recharge. The infiltration rate was found to be in the order of 8–12 m/d. The tertiary treated wastewater might be considered as a water resource for large scale groundwater recharge projects.

ACKNOWLEDGEMENT

The work presented in this paper was conducted within the activities of the project entitled "Modeling of groundwater in two selected coastal aquifers of UAE and Oman as a precursor for Water Resources Management". The project is conducted within the framework of the collaborative research agreement between UAE University and Sultan Qaboos University. The Sharjah Electricity and Water Authority (SEWA), UAE has allocated all the needed resources to conduct the recharge experiment and provided a water supply line from one of its fields. The Ministry of Agriculture and Fisheries, UAE has provided many useful information and data.

SELECTED REFERENCES

Electrowatt Engineering Services Ltd., 1980. Wadi Ham Dam for Groundwater Recharge and Loss Prevention, Preliminary Report, Ministry of Agriculture and Fisheries, UAE.

Electrowatt Engineering Services Ltd., 1981. Wadi Ham Dam and Groundwater Recharge Facilities, Volume I Design, Ministry of Agriculture and Fisheries, UAE.

Entec Europe Limited (ENTEC), 1996. Survey on Groundwater Recharge and Flow in Wadi Ham. Volume 1, Ministry of Agriculture and Fisheries, UAE.

Economics and Social Commission for Western Asia (ESCWA), 1997. Advisory services to the Ministry of Agriculture and Fisheries, United Arab Emirates, Mathematical of the Wadi Ham aquifer, Fujierah coastal plain, UAE.

IWACO 1986. Ground water study. Drilling of Deep Water Wells at Various locations in the UAE.
Geoconsult and Bin Ham (1985). Project 21/81. Drilling of Deep wells at various locations in the UAE. Internal Report (vol. 5). Groundwater Development in the Northern Agricultural Region. Ministry of Agricultre and Fisheries, Dubai, UAE.
Sir Williams Halcrow & Partners Consulting Engineers, 1969. Water Resources of Trucial States. Internal report. Ministry of Agriculture and Fisheries, Dubai, UAE.

Arid Land Hydrogeology: In Search of a Solution to a Threatened Resource – Mohamed(ed.)
© 2006 Taylor & Francis Group, London, ISBN 0 415 41127 0

Sustainable use of groundwater in highly populated areas of the coastal belt of Sri Lanka

M.F. Nawas
Centre for Soil & Water Research, South Eastern University of Sri Lanka, Sammanthurai, Sri Lanka

M.I.M. Mowjood & L.W. Galagedara
Department of Agricultural Engineering, University of Peradeniya, Peradeniya, Sri Lanka

ABSTRACT: Groundwater is the main water source for drinking and other domestic purposes in the eastern coast of Sri Lanka. To assess the sustainability of the safe drinking water 26 representative wells, from a highly populated village named Sainthamaruthu, were selected and the basic water quality parameters were monitored for one year. Height of water table and rainfall were too measured during the period. The water table varied by 2 m between wet and dry seasons. Variations in quality parameters too were observed between the two seasons. Fresh water availability is decreasing with increasing population growth due to increased consumption and pollution. Although, rainfall recharges the aquifers sufficiently high population density, lack of proper drainage systems and improper excreta disposal methods are some factors directly affecting the shallow aquifers, hence groundwater quality subsequently the quantity available. Sustainability, therefore, depends on maintaining the efficient sanitation condition of area.

1 INTRODUCTION

Water supply and sanitation activities have shown remarkable progress in Sri Lanka during last several decades. The government is aiming to provide access to safe drinking water by 2010. However at present, about 75% of urban population and only 14% of rural population are served by pipe borne water. It should be noted that considerable number of people do not have access to those services and large-scale needs may exist in northern and eastern parts of the country.

Population increase creates high demand for the fresh water in terms of quantity and quality. Fast population growth with accelerated urbanization, combined with scare water supplies and poor sanitation, means that governments often cannot supply enough water to meet demand (WHO, 2000). A major percentage of fresh water is stored in aquifers as groundwater resources. Groundwater is the traditional potable water resource for about 70% Sri Lankans (Gunawardana, 2001), and the main source of drinking water especially in the coastal belt of northern, eastern and most part of southern provinces. As such the groundwater resource is very significant in the overall management of the fresh water resources in the coastal belt for its sustainable use.

In many important aspect of public life the term 'quantity' is always given a predominant position whereas the term 'quality' applicable to the same aspect is either ignored or not being given adequate consideration in Sri Lanka. This is true for water which is the most important natural resource vital for life. However, some consideration is given for the quality of 'potable' water since there is an obvious undesirable effect of the potable water of poor quality.

The area chosen for study Sainthamarudhu is a small town in the coastal belt of the eastern province of Sri Lanka. It is bounded by the sea on the eastern flank and by paddy fields and a river on its western flank and is densely populated with 24,069 people residing in 6,266 house-holds (DS Sainthamarudhu, 2003) all of them are confined to an approximately one square kilometre.

Although highly populated areas in Sri Lanka are provided with pipe borne water which is treated and guaranteed by the National Water Supply and Drainage Board (NWSDB) for domestic consumption, this facility is neither adequately provided nor attracted by the general public in Sainthamarudhu. This is mainly due to the availability and accessibility of groundwater as an alternative cheap source of water. This tendency implies that the public is not fully aware of the consequences of continuously increasing high population density on the sustainability of safe groundwater. Only around 15% of pipe borne water is supplied to the entire Ampara district by NWS & DB due to various reasons such as availability of water source, financial constraints, personal preferences, *etc*. The number of pipe borne water supply connections provided to Sainthamaruthu was 857 out of the total of 6,266 families or house-holds while a total of 4,428 private or common wells have been used for drinking and/or other purposes (DS Sainthamarudhu, 2003).

It is noted in this area that most of the houses have their own well and a toilet with a septic tank. These wells and septic tanks are located regardless of the proximity to neighbouring wells and/or septic tanks, especially due to the limited land area for each dwelling. According to Larry and Knox (1985), in highly populated urban and rural areas, due to poor locations for many septic systems, as well as poor designs and constructions and maintenance practices, septic systems have polluted, or have the potential of pollute, underlying groundwater. Areas with more than 40 septic tank systems per square mile can be considered to have potential contamination problems. There were 5,432 septic tanks present in Sainthamaruthu in the same residential area of one square kilometre (DS Sainthamarudhu, 2003). Aquifer of this area belongs to the type of 'shallow aquifers on raised beaches' (Panabokke and Sakthivadivel, 2002). According to Nawas, *et al* (2005), the soils are extremely permeable, leaching significant quantises of both chemical and biological pollutants and therefore the groundwater is highly vulnerable for nitrate and faecal contaminations. Salt water intrusion is another threat for the sustainability of the fresh water.

Information and study results on various aspects of groundwater of north and north-western coast are available in Sri Lanka. However, only a few preliminary studies have been carried out in the south-eastern coast of Sri Lanka. This study was carried out, therefore, to understand the impact of water quality on sustainability with the following objectives: (i) to assess the availability of safe groundwater (ii) to determine the causes of groundwater contamination and (iii) to identify possible conservation methods for the sustainable use.

2 MATERIALS AND METHOD

Sri Lanka, which is an island country with average annual rainfall well over 2500 mm, is located in the south-eastern corner of India. However the rainfall distribution within the country, both spatially and temporally, is not uniform. Vast area of the country comes under dry zone which includes the entire eastern coastal region. Sainthamaruthu is located within the Kalmunai municipal limits/Ampara district of the eastern province of Sri Lanka (Figure 1). The topography is slightly undulating and the highest and lowest points are 3.93 m and 2.39 m, respectively above the mean sea level. The average annual rainfall is around 1900 mm and concentrated between the months of October and January.

Twenty six open dug wells were selected for monitoring of the water quality, locations shown in Figure 1. The study was conducted for nearly one-year period from July 2004 to June 2005. The water table of each selected well was measured in every fortnight as depth to the water surface from the ground level using a wooden stick and a measuring tape.

Wells were sampled in two weeks interval and tested for pH, salinity measured as electrical conductivity, nitrate concentration and faecal pollution measured in terms of *Escherichia coli* as indicator bacteria whose presence in water implies some degree of biological contamination. Electrical conductivity and pH were measured at the site itself by using a portable pH and conductivity meters (Cond 315i, VWR). Water samples were collected into clean (labelled) bottles and preserved in a Styrofoam box until transported to the Centre for Soil and Water Research, South Eastern University of Sri Lanka, where they were analysed for nitrate concentration and *Escherichia coli* enumerations.

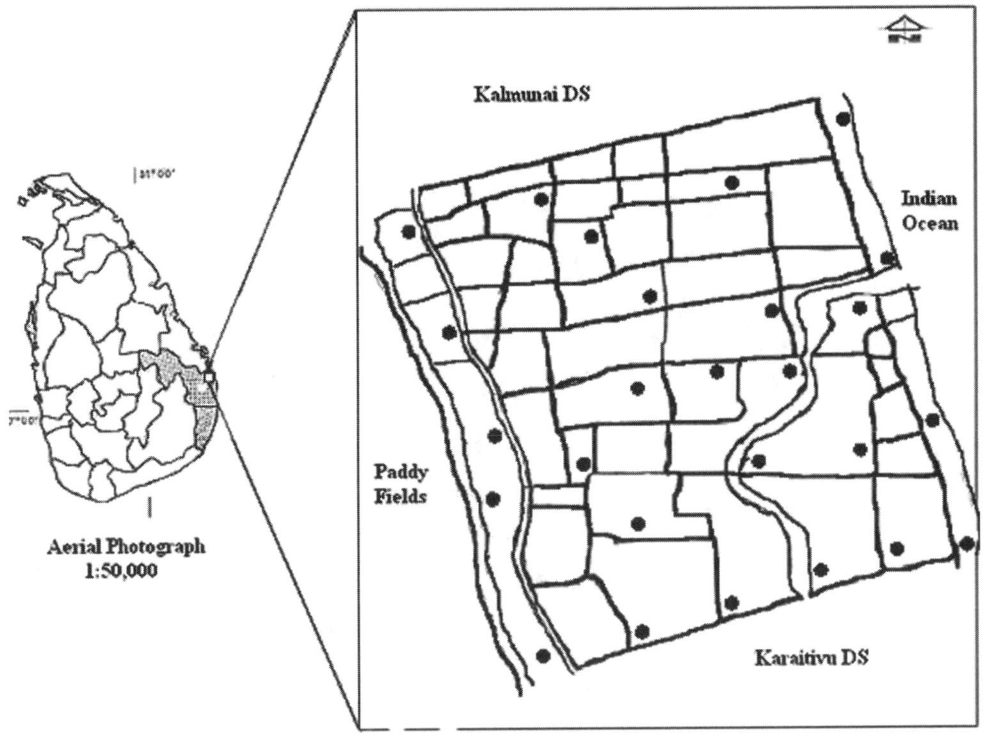

Figure 1. Location of the study area Sainthamaruthu DS and the distribution of observation wells.

Membrane filter technique was used to enumerate the *Escherichia coli* in each water sample. Nitrate concentration in the water samples was measured in the laboratory by means of Spectrophotometer (DR 2010, HACH) using the cadmium reduction (powder pillow) method (HACH, 2002).

3 RESULTS AND DISCUSSION

3.1 *Seasonal variability of the groundwater*

The temporal rainfall distribution in the south-eastern region of Sri Lanka (Figure 2) shows a large variation in a calendar year with at least three months record no rainfall. The mean annual rainfall of 1,902.0 mm is mostly due to intense rains of north-east monsoon (October–January). The monthly rainfall patterns have also shown considerable variation over the years (Figure 2). Although the mean annual rainfall and the consequent recharge of aquifer within the basin is reasonably good, the area experienced a serious shortage of groundwater during the period following no rainfall (Figure 3) in the recent past causing immense hardship to the inhabitants of this area. According to De Silva *et al*, (1999) recharge of the aquifers in the dry zone of Sri Lanka mainly takes place only during the wet season.

During the study period, the total rainfall received was 2,304.3 mm, while the maximum monthly total rainfall of 872 mm was recorded in December 2004 and the minimum monthly total rainfall of 0.0 mm was recorded in June 2004 & 2005.

According to Nawas *et al*, (2005) the quality of water depends on the seasons. During the wet season the microbiological quality, in terms of indicator bacteria *Escherichia coli,* was found poor, while chemical and physical parameters such as nitrate concentration and salinity, in terms of electrical conductivity seems to be within norms in majority of the wells.

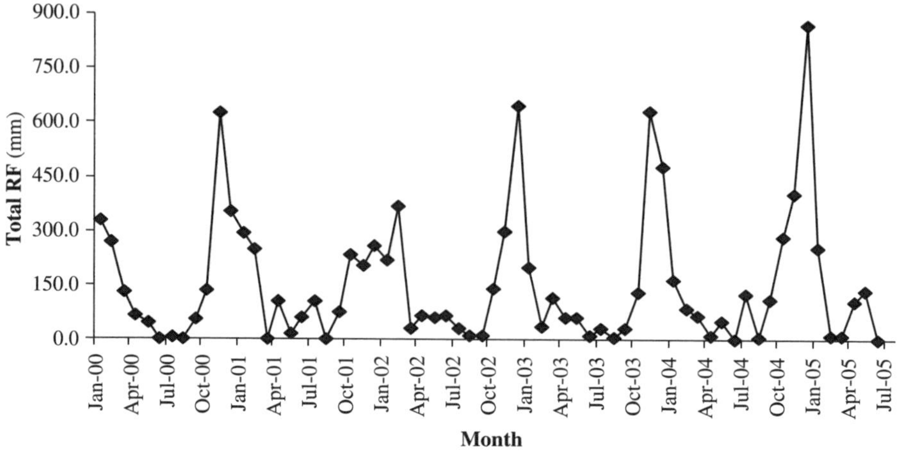

Figure 2. Temporal variation of rainfall in the south-eastern region.

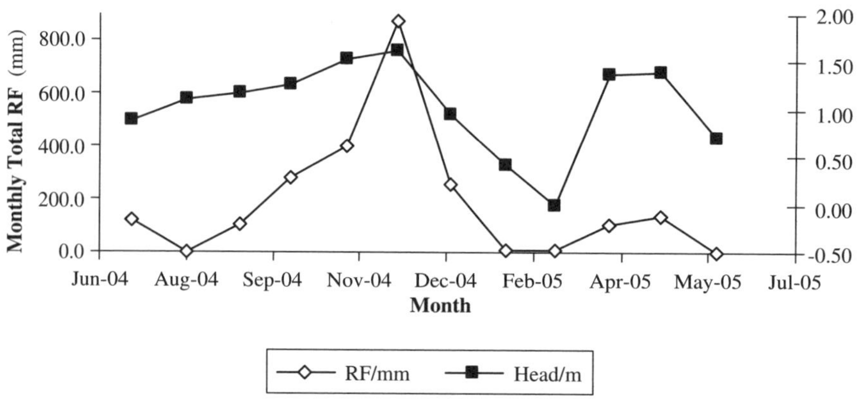

Figure 3. Influence of rainfall on water head in a well.

The cause for faecal contamination, measured in terms of *Escherichia coli*, is mainly due to the high density of septic tanks located within the same small residential area, regardless of their radial distance with nearby dug well. Although, the recommended distance (SLS 745: 1986) between a septic tank and a dug well is 15 m, Figure 4 illustrates that the rule is overlooked, may be due to space constrains. Here the problem of lateral seepage is minimal, because almost all the wells and septic tanks are reinforced with concrete. Wells are elevated around 1 m above the ground level, eliminates the runoff water entering into them. However, the depth is the factor as the water table rises with rainfall; contamination is possible due to elevated water table especially during wet seasons.

All individual wells show variation of electrical conductivity value for rainfall less than 300 mm, but the value remains fairly constant for higher rainfall periods. Comparing the spatial distribution of wells, the best quality water is available in the area close to the main road (far from sea). Seasonal variation of electrical conductivity, shown in Figure 5 clearly indicates the salinity hazard especially during the dry season. The water in wells numbered 8, 13, 14, 20, 21, 22 & 23 showed higher electrical conductivity values, both in wet and dry seasons, due to its proximity to the sea, may be causing sea water intrusion hence higher salinity. Also, it was found that water levels (Figure 6) were below the mean sea level near the sea. Hence, there is a high hydraulic potential

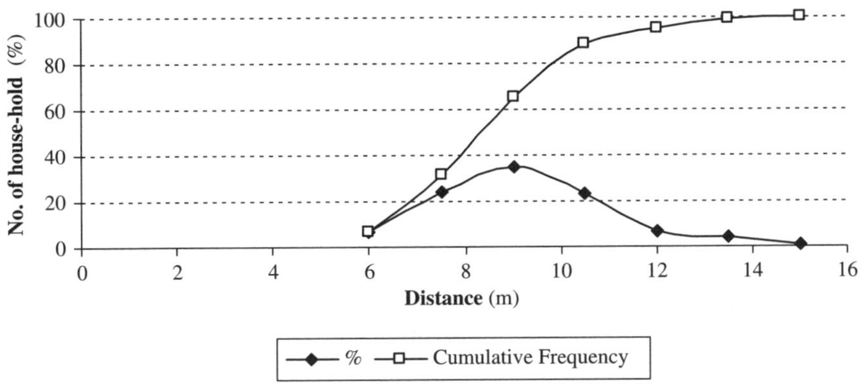

Figure 4. Radial distance between dug wells and septic tanks.

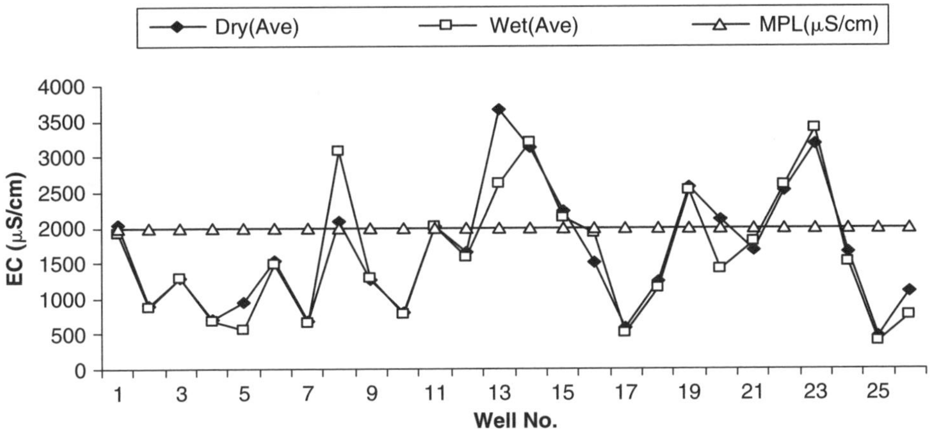

Figure 5. Spatial distribution of average electrical conductivity (EC).

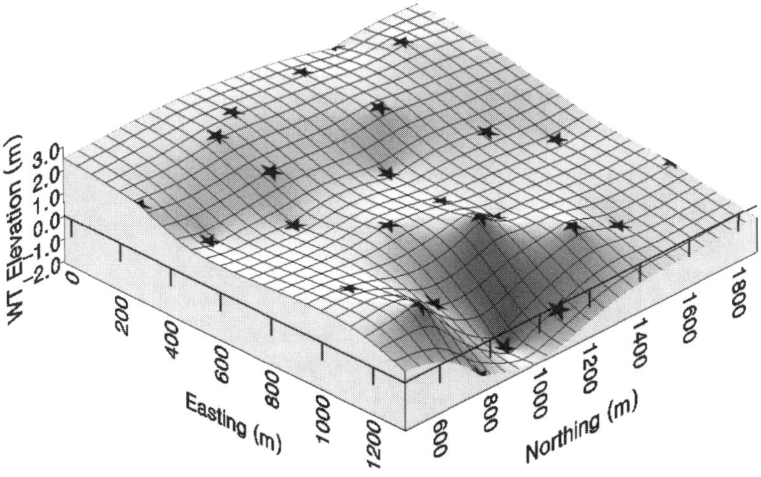

Figure 6. Spatial variation of the water table within the study area with respect to mean sea level.

for water to flow from sea to those well that are located within the coastal area resulting high salinity. However, in most of the wells electrical conductivity values were under the desirable limit but unlimited exploration of groundwater, especially from coastal areas need immediate attention.

4 CONCLUSION

The quality, as well as quantity of water in most of wells from north-east to south-west coastal region was fairly good except in a few isolated areas, where water contamination and low yield were reported. Since people are taking water from the main aquifers at a rate significantly faster than the rate of water recharge, groundwater has been depleting and water scarcity arises during dry seasons as the aquifers are mainly recharged by rainfall. Contamination of the groundwater is a serious problem, which seems to be the major threat for sustainable use. Nitrate contamination is alarming and mostly due to septic tanks and nearby agriculture activities. Salinity is mostly observed in wells that are closer to the sea, possibly due to sea water instruction within 100 m from sea. Unlimited exploration of groundwater from coastal area needs immediate attention. Faecal contamination is widely observed and pose the major threat to the quality and hence sustainability of groundwater. This is due to high density of septic tanks, located in the highly permeable sandy soil. Groundwater of eastern coast, in general and Sainthamaruthu in particular, can be considered highly vulnerable for contamination, as there is no mechanism to regulate the use of water or control pollution. This situation results in a decrease in quantity of available safe water posing threat to sustainability in future. Therefore, the public living in this area has to be advised to take necessary protective measures to maintain hygienic conditions against water contamination. Introducing common sewer system and providing potable water supply to all residents would be the solutions to these identified problems.

REFERENCES

DS Sainthamarudhu (2003). DS office report, Sainthamarudhu

Gunawardana, H. D. (2001). Irrigation Water Quality and Management; in Training Seminar "Water Quality Management, A Need in the New Millennium", conducted by Institute of Chemistry, Ceylon at the Colombo Hilton, on May 30, 2001. pp 7–8, 25

HACH (2002). Water Analysis Handbook, 4th Edition, pp 573–578

Larry W. Canter and R. C. Knox (1985). Septic Tank System Effects on Ground Water Quality. CRC Press. pp 1–3

Nawas, M. F., M. I. M. Mowjood and L. W. Galagedara (2005). Contamination of Shallow Dug Wells in Highly Populated Coastal Sand Aquifer: A Case Study in Sainthamaruthu, Sri Lanka. Tropical Agricultural Research Vol. 17: 114–124

Panabokke, C. R., and Sakthivadivel, R. (2002). Coastal sand aquifers in Sri Lanka, general characteristics, present utilization and gaps in knowledge. Paper presented at March Workshop on 'Critical Issues Facing Groundwater Section in Sri Lanka and south India'. Kodaikkanal, Tamil Nadu pp 1–20.

SLS (1986). Sri Lanka Standard Institutes, 745

WHO (2000). Global water supply and sanitation assessment report, 79. (Available at: <http://www.who.int/water_sanitation_health/Globassessment/Glasspdf TOC htm>

Arid Land Hydrogeology: In Search of a Solution to a Threatened Resource – Mohamed(ed.)
© 2006 Taylor & Francis Group, London, ISBN 0 415 41127 0

Risk management of polluted water bodies

A.M.O. Mohamed
Department of Civil and Environmental Engineering UAE University, UAE

ABSTRACT: It is apparent that if sustainable development is to be achieved, and the requirements of the environmental impact assessment process satisfied, environmental and social considerations as well as technical and economic criteria must be considered from the very beginning of the planning process of any engineering project. This requires leadership on the part of the engineering profession, since it has traditionally dominated the planning process, as well as an evolution towards a more multi-disciplinary approach to design that involves the input of biologists, sociologists, anthropologists, etc., at the earliest stages.

The case study presented in this paper is utilized to develop and evaluate risk management scenarios applicable to a specific polluted surface water bodies. The goal of the case study is to integrate a set of technologies that are capable of eliminating the site's risks to the degree required by pertinent government standards. The set of risk management options must satisfy the sustainable development criteria discussed by Mohamed (1999 & 1998). The method employed is systematic, and demands multiple inputs from various disciplines.

1 INTRODUCTION

1.1 *History of the site*

The canal was constructed to facilitate navigation between various cities. Over the years, various industries were established on the banks of the canal. Presently, at a number of places, the presence of pollution bears witness to a history of industrial occupation. For many years, discharges from the sanitary sewer systems of towns bordering the canal, and industrial emissions, have combined to contribute to the bacteriological and chemical pollution of the waters and sediments of the canal.

A number of years ago, public access to the canal was prohibited, pollution having reached a level judged dangerous to human health and the environment. At that time, certain users of the canal were observed to have contracted skin infections, presumably caused by the presence of pathogenic bacteria in the water. Shortly afterwards, studies were undertaken to determine the nature of the problem. A few years later, the decision was made to cleanup the site.

1.2 *Water quality*

The canal is fed by waters from a large river. The water quality of the river is generally better than that of the canal. For most of the parameters analysed at each station along the canal, the average values are either below the State water quality criteria for recreational activities and the protection of aquatic life (MENVIQ, 1990), or below the detection limit, as shown in Table 1. Only the concentrations of total phosphorus and of faecal coliforms occasionally exceed the quality criteria for the practice of recreational activities with primary contact (swimming, windsurfing, etc). Chromium levels exceeding the criteria for the protection of aquatic life have been found in the water. The average concentrations of copper, lead and total phosphorus exceed the State criteria in the downstream of

Table 1. Water quality in the river, upstream and downstream of the canal.

Parameter	Recreational activities cleanup criteria (mg/l)	Aquatic life cleanup criteria (mg/l)	River (mg/l)	Upstream of the canal (mg/l)	Downstream of the canal (mg/l)
Arsenic	0.05	0.05	<0.001	<0.001	<0.001
Cadmium	0.01	0.0002	<0.001	<0.001	<0.001
Chromium	0.05	0.002	0.004	0.004	<0.003
Copper	1.0	0.002	0.003	0.003	0.0025
Cyanide	0.2	0.005	<0.005	<0.005	<0.005
Mercury	0.001	0.000006	<0.0002	<0.0002	<0.0002
Nickel	0.25	0.025	<0.002	<0.002	<0.002
Lead	0.05	0.001	<0.002	<0.05	<0.002
Zinc	5.0	0.03	<0.05	<0.05	<0.05
Nitrogen	–	1.2	<0.1	<0.1	<0.1
TOC	–	–	7.1	7.5	8.2
Coliform	200.0	–	55.0	203.0	96.0
Conductivity	–	–	77.5	71.0	72.0
SM	–	–	<5.0	21.0	8.0
DO	–	>6.5	12.5	12.0	13.1
pH	–	6.5–9	7.8	7.8	7.7
Phosphorus	5–9	0.03	<0.05	0.09	0.1
PCBs	0.03	0.000001	<0.0001	<0.0001	<0.0001
PAHs	–	–	<0.0002	<0.0002	<0.0002
MO&G	–	–	0.28	–	0.26
Phenol	0.001	0.005	<0.002	<0.002	0.003

TOC = total organic carbon, SM = suspended matter, DO = dissolved oxygen, PCB = poly chloronated biphenols, PAH = poly aromatic hydrocarbons, conductivity (micro S/cm), MO&G = mineral oil and grease, coliform (n/100 ml), cleanup criteria as stated by MENVIQ (1990).

the canal. The observed values are, however, only slightly above the criteria and, consequently, should not affect most aquatic organisms.

Water in the upstream of the canal contains concentrations of cadmium, lead and copper equal to or slightly above the criteria for protection of aquatic life. Although they occasionally reach relatively high values, mineral oils and grease and phosphorus vary randomly throughout the canal. Moreover, the faecal coliforms are higher in the upstream of the canal (203/100 ml) than in the downstream (96/100 ml). The quantity of suspended solids reaches 21 mg/l in the upstream and 8 mg/l in the downstream.

1.3 *Sediment quality*

The industrial activity developed along the banks of the canal has left its mark in the form of a layer of contaminated sediments. In the upstream, most of the pollutants have concentrations below the harmful threshold (Level 3) defined by the State criteria (MENVIQ, 1990), as shown in Table 2. Nevertheless, despite the fact that occasionally the concentrations of certain pollutants are high, the degree of pollution of the sediments in the upstream is relatively low. Consequently, the sediments in the upstream do not appear to be sufficiently polluted to justify corrective measures in the short term.

The results show that the sediments in the downstream of the canal are rather polluted, as shown in Table 3. The concentrations of several of the variables analysed exceed the threshold for harmful effects (Level 3) (MENVIQ, 1990) at various sampling sections. Moreover, analysis of boring samples indicated that the average thickness of the polluted sediment is 260 mm, which results in a contaminated sediment volume of about 120,000 m^3.

Table 2. Sediment quality in the upstream of the canal.

Parameter	Average concentration (mg/l)	Cleanup criteria Level 1 (mg/l)	Cleanup criteria Level 2 (mg/l)	Cleanup criteria Level 3 (mg/l)
Arsenic	2.8	3	7	17
Cadmium	2.1	0.2	0.9	3
Chromium	61.0	55	55	100
Copper	62.0	28	28	86
Mercury	0.51	0.05	0.2	1
Nickel	39.0	35	35	61
Lead	142.0	23	42	170
Zinc	240.0	100	150	540
PCBs	0.12	0.02	0.2	1
MO&G	1407.0	1000	2000	–
TOC (%)	5.2	–	1	10
PAHs	<D L	1	–	–
Nitrogen	5357.0	–	–	–
Phosphorus	1962.0	–	600	2000
Coliform	99.0	–	–	–

Table 3. Sediment quality in the downstream of the canal.

Parameter	Average concentration (mg/l)	Cleanup criteria Level 1 (mg/l)	Cleanup criteria Level 2 (mg/l)	Cleanup criteria Level 3 (mg/l)
Arsenic	5.3	3	7	17
Cadmium	2.4	0.2	0.9	3
Chromium	56.0	55	55	100
Copper	84.0	28	28	86
Mercury	1.72	0.05	0.2	1
Nickel	53.0	35	35	61
Lead	304.0	23	42	170
Zinc	1340.0	100	150	540
PCBs	0.76	0.02	0.2	1
MO&G	6.92	1000	2000	–
TOC (%)	9.9	–	1	10
PAHs	50.2	1	–	–
Nitrogen	2579.0	–	–	–
Phosphorus	1516.0	–	600	2000
Coliform	60.0	–	–	–

Levels 1, 2, and 3 correspond to threshold concentrations for no effect, weak effect, and harmful effect, respectively (MENVIQ, 1990).

2 DEVELOPMENT OF RISK MANAGEMENT ALTERNATIVES

The following cleanup alternatives have been identified for the case study:

(1) Storage of sediments on land;
(2) On-site storage of sediments on the canal bed;
(3) Sediment encapsulation within the canal area;
(4) Off-site solidification/stabilization;
(5) On-site solidification/stabilization; and
(6) Extraction-based treatment of polluted sediments.

2.1 *Alternative 1: Storage of sediments on land*

The exterior storage alternative for canal sediment comprises four major steps:

(1) *Excavation*: The only practical method of removing contaminated sediment from the bed of the canal is dredging. Hydraulic and pneumatic dredges are the best techniques from the technical and environmental viewpoints.
(2) *Volume Reduction*: Reduction of mud (water and sediments) volume could be achieved via separation based on particle size and dehydration.
(3) *Transportation*: Transportation modes depend principally on the characteristics of the dredged materials, the stage at which volume reduction is carried out, and the type of treatment. Pipelines could be used for liquid sediment transport, and trucks for transport of dehydrated sediments.
(4) *External Storage of Polluted Sediment*: The principal storage alternatives are storage on land in existing cells which are managed by various waste management companies (e.g., Cintec Environmental Inc., Quebec; Laidlaw Environmental Services, Ontario; and Chemical Waste Management (CWM) Inc., New York State), and storage on land in new cells which require building maximum security cells according to Provincial and State regulations. Cost estimates show an investment of about Can$28 million for storage at Cintec, about Can$22 million for the CWM site and about Can$5 million for a new cell to contain 120,000 m^3 of polluted sediment.

2.2 *Alternative 2: On-site storage of sediment on the canal bed*

The on-site method of polluted sediment stabilization would consist of storing the sediment on the canal bed, covering it with a permeable geotextile membrane, then compacting it by overlaying a bed of crushed stone on top of the membrane. The estimated cost to carry out this option is Can$6 million.

2.3 *Alternative 3: Sediment encapsulation within the canal area*

Encapsulation involves storage of polluted sediment in watertight holding basins constructed within the canal itself. Polluted sediments would be excavated by means of suction flow-reversing dredges and transferred to the holding sections by means of pipelines. The holding sections would act as sedimentation-settling zones, allowing only water that meets the established quality standards to return to the canal. Each watertight section would be subdivided into two compartments for a progressive treatment of the water-sediment mixture. The upstream compartment would receive the water-sediment mixture from the dredging activity for primary decantation, which could be accelerated by adding coagulating agents. It would be isolated from the canal and the downstream compartment by means of watertight geomembranes. The bed and walls of the canal and the lower half of the dike of the downstream compartment would be isolated by watertight geomembranes while the upper half of the dike would be covered by a geomembrane filter. Following treatment, water could be returned to the canal through an outlet located in the downstream part of the compartment, or through the geomembrane filter.

When the sediment has completely settled, the two compartments would be covered by a semi-permeable membrane, allowing the escape of gas produced by the decomposition of organic matter in the sediment. A protective layer of sand or any type of clean material would be used as a cover. It will, also, compact the sediment beneath the membrane. The space occupied by the compartments would reduce the width of the canal from 50 m to 20 m, which would still leave sufficient space for the practice of aquatic activities. The total estimated cost for implementing this option is about Can$10 million.

2.4 *Alternative 4: Off-site solidification/stabilization*

Solidification/stabilization, as discussed by Mohamed and Antia (1998), is a treatment technique already in use in various parts of the world. The technique is suitable for treatment of heavy metals and oils and greases at low concentrations. The highly alkaline pH (approx. 12) of treated material prevents any re-colonization by microorganisms. In addition, pathogenic organisms are killed during this treatment.

Off-site treatment requires prior dredging of polluted sediments. A sediment storage zone and a treatment zone are then setup on the banks of the canal. Pretreatment could be necessary for material preparation before adding the solidifying agents. Pretreatment may consist of filtering to remove large particles, breaking up aggregates in order to homogenize the material, or addition of additives to improve the sediment texture. The necessity for pretreatment, together with its nature, is evaluated following characterization of the sediments to be treated. Next, chemical reagents are added to the polluted sediments in a mixer, or possibly an ordinary cement truck or mixers (reservoirs equipped with agitators) specially installed on-site during operations. Concentration of reagents to be added to the polluted sediments is evaluated based on sediment characteristics. Water is also added to the mixture, for hydration and chemical reactions. After mixing, sediments and reagents form a product which has the consistency of liquid cement. It is then placed in cells where it may undergo a maturation phase. The period of maturation varies from 24 hours to several months, depending on the reagents added, and the future use of the treated sediments.

Several schemes for final disposal may be considered for the product of off-site solidification/stabilization treatment. The treated material could be sent to landfills, where it could be used as covering material. It could also be used as embankment material or for general construction purposes. Whatever the final use may be, the treated material should be in conformity with regulations and guidelines for solid waste disposal practice. The unit cost for off-site treatment varies between Can\$50 and Can\$100 per m^3 of polluted sediments. Hence, treatment of 120,000 m^3 of sediments would result in total costs varying from Can\$6 million to Can\$12 million.

2.5 *Alternative 5: On-site solidification/stabilization*

On-site solidification/stabilization is accomplished by adding directly to sediment (i.e., in their location in the canal) a chemical reagent, the proportion of which will vary according to the degree of contamination. Cement is added next in order to produce a highly resistant material with low hydraulic conductivity. The maturation and solidification period of the sediment/reagent mixture occurs underwater within several hours. Since treatment occurs underwater, without prior dredging of sediment, no aqueous effluent is generated. The work zone for on-site solidification/stabilization is the canal itself. Equipment used includes the drilling and mixing system, a crane, tanks that contain reagents, computerized equipment for process control and generators, all of which are located on a barge. The treated material can be left in place or be used to reinforce the sidewalls of the canal provided that it conforms with the solid waste disposal regulations. The unit cost of on-site treatment varies between Can\$100 and Can\$200 per m^3 of polluted sediments. Thus, the treatment of 120,000 m^3 of sediment would result in total costs varying between Can\$12 million and Can\$24 million.

2.6 *Alternative 6: Extraction-based treatment*

Extraction-based treatment comprises several stages. First, the sediments are washed (physically extracted) in order to separate the fine particles in which pollutants are concentrated. Next, metals contained in the fine portion are chemically extracted and recuperated via solvent extraction techniques, as discussed by Mohamed and Antia (1998). Finally, organic pollutants are either treated via surfactant extraction, bioremediated, landfilled or vitrified, etc. Due to the nature of pollutants, organic and inorganic, it is necessary to follow a consistent sequence of washing, metal extraction and treatment of the organic pollutants.

2.6.1 *Sediment washing*

Sediment washing is a physical operation, consisting of separating the fine particles from the coarser particles of the sediment. The need for particle separation is based on the hypothesis that pollutants are attached to the finer particles of the sediment. Pollutants, also, attach to soil organic matter, notably due to its high adsorption capacity (Mohamed & Antia 1998, Yong et al. 1992). Coarse particles, i.e., more than approx. 5 to 10 mm diameter, are assumed not to be contaminated. Following an initial separation, subsequent stages of particle separation are carried out, namely, separation of particles by centrifugal force, metal extraction (e.g., electromagnetic and electrostatic processes), etc.

Sediment washing results in coarse particles of clean sediment, and fine particles in which metals and organic pollutants are concentrated. The process also generates polluted water which must be treated before discharge. Air purifiers such as active carbon filters or biofilters should be used.

Once particles of diameter exceeding 5–10 mm are removed, the sediment may be subjected to a second type of treatment, i.e., high pressure washing. This type of washing keeps sediments in contact with water under high pressure, gradually forcing separation of fines from coarse particles. The larger washed particles are considered clean whereas the finer particles contain both organic and inorganic pollutants. High pressure washing permits volatilization of certain volatile pollutants, which either remain in the gaseous phase or dissolve in the aqueous phase. Air and water purification systems should, therefore, be in place during this type of treatment.

2.6.2 *Metal extraction*

Following the washing stage, the fine particles of sediment, containing concentrated organic and inorganic pollutants, are subjected to a second treatment stage, namely, metal extraction. The principle of metal extraction, as discussed by Mohamed and Antia [3], consists of first separating metals present in the fine particles by means of solvents and chelating agents. Next, chelates are separated from the fine particles, which contain only organic chemicals. Metals are then recovered from chelates by using an acid solution to break the chemical bonds between the chelating agents and metals. Then, chelates are regenerated and re-utilized at the upstream of the extraction process while metals are recuperated for recycling purposes by means of electroplating.

2.6.3 *Treatment of organic chemicals*

The last stage of sediment treatment is the treatment of organic chemicals (e.g., PCB, PAHs) that are concentrated in fine particles and water. Several options are available, including surfactant-based washing, biodegradation, and thermal destruction. Incineration is a proven technology, but it is subject to some social resistance. When this technology is used, sediment loses its structure and cannot be reused for any purpose other than land disposal. Biodegradation enables regeneration of sediment and produces only aqueous effluent as residue. Surfactant-based washing or biological treatment of the residue allows recuperation of water for reuse in the process. Nonetheless, the presence of PCBs hinders conventional methods of biodegradation. In this case, dechlorination would be necessary; it could be achieved by irradiation of contaminated particles with ultraviolet rays, which provoke the removal of chlorates from organic molecules. Next, normal decontamination in which chlorine atoms react with an alkaline metal (e.g., Na) to form salts (e.g., NaCl) could be carried out.

Finally, it is possible to landfill, without treatment, organically polluted fine particles. Due to the nature of the contamination, disposal would take place in storage cells or in landfills. In this case, the sediment resource is not reusable. Costs of complete sediment treatment would be between Can$ 250 and Can$399 per m^3 for a total cost varying between Can$30 and Can$43 million for the treatment of 120,000 m^3 of polluted sediments.

3 COMPARATIVE ANALYSIS OF RISK MANAGEMENT ALTERNATIVES

In the present analysis, *environmental*, *technical* and *economic criteria* are used (Mohamed 1998 & 1999). Each criterion is divided into various options and each option is divided into three classes according to their importance. The method consists of rating each possible option according to the criterion. The chosen option is the one that accumulates the most first ratings. If two or more options have the same number of first ratings, a count of second rating is taken. If necessary, the count is continued (third rating, etc.) until the option with the best performance is determined.

The criteria used in this comparative analysis are: (1) permanent environmental, (2) temporary environment, (3) technical, (4) technical and economical, (5) economic, (6) non-discriminatory, and (7) non-pertinent. These criteria are discussed below.

3.1 *Permanent environmental criteria*

Option 1: Elimination of pollution in canal sediment
This option is used to evaluate whether an option is able to decontaminate the canal sediments. Treatment options eliminate pollution. In containment and fixation options, pollutants are still present, although they are immobilized. The criterion reflects a concern that the sediment pollution problem should, if possible, be permanently resolved. This means that options which ensure permanent sediment cleanup should be preferred over others. Possible ratings are shown in Table 4.

Option 2: Sediment management in study zone
This option expresses the desire to resolve the contamination problem in the study zone. Transfer of contamination elsewhere could cause social opposition (Not in My Back Yard, NIMBY, syndrome) which could compromise the feasibility of the project. Possible ratings are shown in Table 4.

Option 3: Heritage
This option evaluates changes in the layout of the canal, since it is preferable to preserve the historic or present layout of the canal. Possible ratings are shown in Table 4.

Option 4: Risks of polluting groundwater and surface water after completion of treatment
This option covers risks of contamination of the canal and groundwater in the study zone following decontamination operations. Risk of future pollution is directly linked to the presence of pollutants and the possibility that they may again be found in the water. Possible ratings are shown in Table 4.

Option 5: Substratum characteristics following completion of work
Once sediment has been treated or removed, the final characteristics of the substratum at the bottom of the canal will influence re-colonization of the canal by flora and fauna. Possible ratings are shown in Table 4.

Table 4. Possible ratings of options of permanent environmental criteria for the comparative analysis.

Option	Rating	Justification
1	1st	1st Complete treatment and elimination of pollutants in sediments
	2nd	Confinement or fixation of polluted sediments
2	1st	Sediments remain in the study zone
	2nd	Part or all the sediments is transferred outside the study zone
3	1st	No changes in the layout of the canal
	2nd	Changes in the layout of the canal
4	1st	No pollution risk; contaminants are no longer in the study zone
	2nd	Low risk of pollution; contaminants confined or fixed in the canal in a completely closed compartment
	3rd	Moderate risk of pollution; semi-closed compartment
5	1st	Substratum similar to natural state
	2nd	Substratum composed of small stones (added material)
	3rd	Cement substratum

3.2 *Temporary environmental criteria*

Option 6: Inconvenience caused by presence of construction site
This option covers all kinds of inconvenience which may impact the public during work period. Inconvenience includes dust, operational noise arising from the cleanup process, odours, and deterioration of the landscape. Inconvenience connected to closing or diversion of the cycle path in several places, due to the presence of a construction site, is also evaluated, as is the duration of such inconvenience. Inconveniences linked to the presence of a construction site beside the canal will principally affect the population living near the canal, or those frequenting the canal. Possible ratings are shown in Table 5.

Option 7: Inconvenience due to trucking activity
This option covers such matters as perturbation to circulation and pedestrian security. It takes into account the need to access the canal. Inconvenience from trucking activity will affect the population living near the canal, and will also cause repercussions for several kilometres around the work site. Possible ratings are shown in Table 5.

Option 8: Impact of works on flora and fauna
This option is used to evaluate negative impact (injury, mortality) caused by carrying out a decontamination option, such as that arising from lowering the water level or from turbidity due to dredging or hydraulic agitation. Possible ratings are shown in Table 5.

3.3 *Technical criteria*

Option 9: Works linked to substructures after completion
This option evaluates the possibilities of installing new substructures (e.g., subterranean cables), as well as the ease of carrying out and maintaining operations on substructures already in place, after work completion. These substructures include servitudes, subterranean installations, and installations underwater or on the banks of the canal. Possible ratings are shown in Table 6.

Option 10: Technical feasibility
This option covers technological risks linked principally to the stage of development or applicability of technologies selected for decontamination of the canal. Possible ratings are shown in Table 6.

Table 5. Possible ratings of options of the temporary environmental criteria for the comparative analysis.

Option	Rating	Justification
6	1st	Majority of work taking place outside study zone; minimal installations, little storage required, work completed within a year or less
	2nd	Work carried out in study zone; installations on-site, including storage, work completed within a year or less
	3rd	Work carried out in study zone; installations on site, including storage, work completed in more than a year
7	1st	Minimal trucking activity (movement, stopping, loading)
	2nd	High level of trucking activity (movement, stopping, loading and transport of material from outside)
	3rd	Very high level of trucking activity (movement, stopping, loading and transport of sediment to outside study zone)
8	1st	Little or no drop in water level, minimal dredging and mechanical agitation
	2nd	Little or no drop in water level, dredging
	3rd	Major drop in water level

3.4 *Technical and economic criteria*

Option 11: Follow-up measures in study zone and associated costs
This option evaluates the amount of follow-up which will be necessary after completion of decontamination in the study zone. Follow-up consists of ensuring, on a long term basis, the effectiveness of the restoration. Costs involved are proportional to the amount of follow-up required. This criterion also covers the ease with which appropriate follow-up measures can be carried out. Possible ratings are shown in Table 7.

3.5 *Economic criteria*

Option 12: Total cost
This option evaluates the total cost of the project, with regard to each option, for treatment of 120,000 m^3 of polluted sediments. Possible ratings are shown in Table 7.

3.6 *Non-discriminatory criteria*

The following options are deemed non-discriminatory, i.e., do not differentiate between the various options, since their influence on all the options is similar.

Option 13: Contaminant risks downstream from the study zone
The six treatment alternatives under consideration are not discriminated by this option. All the alternatives, except on-site solidification/stabilization, require dredging of polluted sediment, which results in re-suspension of contaminated sediment. However, measures can be provided to reduce this effect. With regard to on-site solidification/stabilization, protective measures are included in the design. Furthermore, the treatment work area could be sealed off by preventing water flow from upstream, thereby preventing downstream pollution. Hence, this option cannot be used to discriminate one treatment alternative from the other.

Table 6. Possible ratings of options of the technical criteria for the comparative analysis.

Option	Rating	Justification
9	1st	Canal bottom similar to natural state
	2nd	Canal bottom similar to natural state but compartments present on canal banks
	3rd	Characteristics of canal bottom altered (e.g., cement or geomembrane)
10	1st	Proven technology for similar treatment
	2nd	Proven technology for a different treatment
	3rd	Technology not commercially proven (at pilot or pre-industrial stage)

Table 7. Possible ratings of options of the technical and economic criteria for the comparative analysis.

Option	Rating	Justification
11	1st	No follow-up required, pollutants are removed from study zone
	2nd	Control is required, but easily carried out, i.e., small number of sampling points is needed for appropriate follow-up
	3rd	Control is required and is difficult to carry out; large number of sampling points is required for appropriate follow-up
12	1st	Can$10 million or less
	2nd	Can$10–30 million
	3rd	over Can$30 million

Option 14: Potential for recreational, faunic and landscape redevelopment
This option is considered non-discriminatory because all treatment alternatives considered will result in re-opening of the canal to the public. The potential result will be an improvement of the environment, leading to potential recreational, faunic and landscape redevelopment, and the effect will be similar no matter which treatment alternative is selected.

Option 15: Impact of works on the local economy
Construction work of any type will have a direct impact on local economy. For example, perturbation of traffic and closing of access roads will adversely affect some local businesses, although an increase in workforce could result in increased business for certain local establishments (e.g., restaurants, etc.). In addition, regarding employment, the call for bids could specify that a percentage of local personnel must be hired for the project, no matter which option is selected. Therefore, this option is considered non-discriminatory.

Option 16: Impact on property values and the socio-economic structure
Speculation stimulated by the decontamination of the canal, no matter which treatment option is utilized, will have similar effects on real estate market.

3.7 *Non-pertinent criteria*

Option 17: Industrial activity during work period
Some of the options involve a reduction, major or otherwise, in water level in the canal during part of the work. This drop in water level would cut off supply to some industries along the side of the canal. However, it must be stipulated in the call for bids that water supply to various industries be assured throughout the work period.

Option 18: Stability of canal walls
If the canal is drained for a long period, measures should be taken to prevent the side walls from caving-in (e.g., technical inspection during installation of metal reinforcement to the wall structure). These requirements would be stipulated in the call for bids, thus making preventive or protective measures of this nature mandatory.

Option 19: Time table
The actual timetable will be known only after tenders are received following call for bids. Once again, a time limit for work completion could be laid down in the call for bids. The expected time required to carry out the work, according to each option, is included in option 6.

3.8 *Importance of rating*

Rating allows a relative weight to be attached to each of the options, those options deemed more important being given a heavier weight. In this analysis, three classes of criteria are defined, the first and third representing the most and least important, respectively.

Since the choice of option depends on the number of top ratings accumulated, the importance of the several options is taken into account by diminishing the ratings given to the options of the less importance. The revised rating given to an option after the importance of the class has been allowed for will be called the weighted rating. For example, a second rating, for a top-rated option, will have the same weight as a first rating for a second class rated option. Both will be given a second-weight rating. As the maximum number of ratings (class of importance) possible is three, a maximum of five corresponding ranks will be possible, as indicated in Table 8 (Holmes 1971).

The class of importance was determined from a perspective of sustainable development. The class of importance assigned to each of the twelve options is shown in Table 9. It is seen that out of the twelve options used in the comparative analysis, five were given first class rating, four a second class and three a third class rating.

Table 8. Corresponding ranks (Holmes 1971).

Class of importance (CI)	Relative rank of option (RRO) First	Second	Third
1	1	2	3
2	2	3	4
3	3	4	5

Table 9. Class of importance for various options.

Option	Class	Rational
1	2	All options conform to the main objective of the project, i.e., "*as far as possible the choice of option should favour permanent treatment of contaminated sediments*."
2	1	This option expresses the State responsibility
3	3	The public has expressed a desire to preserve the present layout of the canal
4	2	Risks of future contamination of canal waters are linked to the presence of pollutants, which could seep to subsurface water in places where the hydraulic gradient is oriented from the canal towards inclined layers. Nonetheless, a follow-up after completion of work would enable problems to be discovered and corrective action to be taken, if necessary
5	3	Changes of Canal's substrate is possible; the rate and diversity of the substrate changes may, however, be influenced by the type of substrata in place
6	1	This option is a matter of social acceptability. Alteration of the surrounding conditions usually means increased noise, dust, and unpleasant odours. This may create negative impacts to the local population. The project could be contested by the population owing to inconveniences during work in progress. However, some measures are possible for reducing the nuisance (e.g., appropriate working hours, communication plan, etc.)
7	1	This option is also a matter of social acceptability. Inconveniences related to urban transportation will directly affect the population, business and adjacent industries. Protests could result from these inconveniences
8	3	This option is important during progress of work. It is temporary. No rare or endangered species are to be found in and around the study zone. Changes are possible even though the canal does not constitute an important environment for wild flora and fauna
9	2	Access to or installation of new substructures on the canal bottom following decontamination will be complicated by a substratum different from the natural one, especially if the new substratum in intended to be watertight. The presence of compartments on the banks could interfere with some future work. Nonetheless, some modifications or adjustments could be made to overcome these difficulties
10	1	Technical feasibility of the various decontamination options implies the notion of technical risks associated with relatively undeveloped or commercially unproved technologies. Attainment of the project objectives depends directly on the decontamination technology selected
11	2	Follow-up after work completion requires long term effort and investment, which must be added to costs already incurred
12	1	The total cost of the project is very important, as it has been determined that the status quo does not involve any public health risk and does not require immediate action. Thus, the public could object if a very costly project was undertaken. On the other hand, if public health was threatened, little importance would have been attached to the total cost of the project, the priority being protection of the population

3.9 *Risk management alternative selection*

Evaluation of each of the six treatment alternatives according to the twelve defined options is discussed previously. Four partial Tables, corresponding to the four subcategories of criteria, i.e., permanent environmental (Table 4), temporary environmental (Table 5), technical (Table 6), and technical and economical (Table 7), are developed.

Ranking of the six treatment alternatives is shown in Table 10. Evaluation of performance of the various options is achieved by calculating the number of first ratings accumulated by each option. If two or more options be awarded the same number of first ratings, the number of second and third ratings should be calculated.

The final ratings of the six treatment alternatives are shown in Table 11. The encapsulation alternative is awarded the highest classification, this being the alternative with the largest number of first ratings. The storage on land alternative is in the second position, followed by chemical extraction. The fourth, fifth and sixth positions are given to on-site solidification/stabilization, storage with geomembrane (FML) and off-site solidification/stabilization.

All methods of comparative analysis have limitations. These limitations are principally attributable to the number and classification of options. Thus, the number of options selected in each category, as well as the classification, is determinant choices which remain subjective, in as much as they are decided upon by a task force comprising a restricted number of specialists. On the other hand, analysis of results of the comparative analysis and most particularly analysis of the strong

Table 10. Ranking of the various treatment alternatives.

		Treatment technique											
		Storage on land		Storage with FML		Encapsulation		On-site S/S		Off-site S/S		Treatment	
Option	Class (CI)	R	RRO	R	RRO	R	RRO	R	RRO	R	RRO	R	RRO
1	2	2nd	3	2nd	3	2nd	3	2nd	3	2nd	3	1st	2
2	1	2nd	2	1st	1	1st	1	2nd	2	1st	1	1st	1
3	3	1st	3	1st	3	2nd	4	1st	3	1st	3	1st	3
4	2	1st	2	3rd	4	2nd	3	1st	2	3rd	4	1st	2
5	3	1st	3	2nd	4	1st	3	1st	3	3rd	5	1st	3
6	1	1st	1	2nd	2	3rd	3	2nd	2	2nd	2	3rd	3
7	1	3rd	3	2nd	2	2nd	2	3rd	3	1st	1	1st	1
8	3	2nd	4	3rd	5	3rd	5	2nd	4	1st	3	2nd	4
9	2	1st	2	2nd	3	2nd	3	1st	2	3rd	4	1st	2
10	1	1st	1	3rd	3	1st	1	2nd	2	2nd	2	3rd	3
11	2	1st	2	3rd	4	2nd	3	1st	2	3rd	4	1st	2
12	1	2nd	2	1st	1	1st	1	3rd	3	2nd	2	3rd	3

Table 11. Final ranking of the various treatment alternatives.

Relative rank of option	Storage on land	Storage with FML	Encapsulation	On-site S/S	Off-site S/S	Treatment
1	2	2	3	0	2	2
2	5	2	1	6	3	4
3	4	4	6	5	3	5
4	1	3	1	1	3	1
5	0	1	1	0	1	0
Ranking	**2**	**5**	**1**	**6**	**4**	**3**

and weak points of the chosen option make it possible to partially overcome the limits of the method and validate its results.

From the results shown in Table 11, the encapsulation option has the highest rating and is therefore, the cleanup alternative selected. Scientific analysis of the individual tables make it possible to point out the strengths and weaknesses of this option. First, it can be noted that encapsulation has a second rating for performance among the permanent environmental criteria. Indeed, sediment management is carried out within the study zone. In addition, the natural granular substratum will remain in place at the end of the project. Encapsulation is a storage process, but the fact that sediment is confined in completely closed compartments limits future risk of water contamination in the canal and groundwater.

In terms of the temporary environmental criteria, encapsulation is ranked last. This constitutes the major weakness of the selected option. For technical performance, encapsulation alternative is rated the second. It is a proven confinement technique which is in commercial use for similar projects. Finally, encapsulation is rated first for economic performance, since this treatment alternative has the least cost.

4 CONCLUSION

It is apparent that if sustainable development is to be achieved, and the requirements of the environmental impact assessment process satisfied, environmental and social considerations as well as technical and economic criteria must be considered from the very beginning of the planning process. This will require leadership on the part of the engineering profession, since it has traditionally dominated the planning process, as well as an evolution towards a more multi-disciplinary approach to design that involves the input of biologists, sociologists, anthropologists, etc., at the earliest stages.

The case study discussed in this paper is utilized to develop and evaluate remediation scenarios applicable to a specific uncontrolled hazardous waste site. The goal of the case study is to integrate a set of technologies that are capable of eliminating the site's risks to the degree required by pertinent government standards. The set of remedial options must satisfy the sustainable development criteria, discussed by Mohamed (1998 & 1999). The method employed is systematic, and demands multiple inputs from various disciplines.

REFERENCES

Mohamed, A.M.O. 1998. "Engineering for Sustainable Development: Development of a Protocol," 2nd International Conference, Role of Engineering Towards Better Environment (Sustainable Development), 12–15 December 1998, Alexandria, Egypt, pp. 1–9.

Mohamed, A.M.O. 1999. "Concept Evaluation and Protocol Development of Sustainable Development for water Resources Planners," 2nd International Conference on Sustainability in Desert Regions (SDR99), 20–22 November, Al Ain, United Arab Emirates.

Mohamed, A.M.O., and Antia, H.E. 1998. "Geoenvironmental Engineering," Elsevier, Amsterdam, p. 727.

Yong, R.N., Mohamed, A.M.O., and Warkentin, B.P. 1992. "Principles of Contaminant Transport in Soils," Elsevier, Amsterdam, p. 327.

Holmes, J.C. 1971. "An Ordinal Method of Evaluation," Urban Studies, Vol. 9, pp. 179–191.

Groundwater assessment

Arid Land Hydrogeology: In Search of a Solution to a Threatened Resource – Mohamed(ed.)
© 2006 Taylor & Francis Group, London, ISBN 0 415 41127 0

Groundwater monitoring evaluation and optimization in arid regions: Abu Dhabi Emirate case study

M.A. Dawoud, M.L. Dash & H.A. Haouqani
Monitoring and Exploration, Water Resources Department, Environment Agency-Abu Dhabi, UAE

ABSTRACT: In an arid region such as the Arabian Gulf, where scare water resources are the norm, the availability of water is a key element in determining the scale of long term development. A well designed water resources monitoring program plays an important role in developing a sustainable and integrated water resources management strategy. Two monitoring programs in Abu Dhabi Emirate commenced in the 1980's and 1990's. The first is the joint NDC/USGS Groundwater Research Project which commenced in 1988. About 600 monitoring wells in 485 locations were drilled and monitored regularly. About 48 wells are equipped with automatic loggers and the others are monitored manually. The second is the joint ADNOC/GTZ-DOC consortium project for groundwater assessment which commenced in 1996. During this project 675 wells were drilled from which 230 are used as continuous monitoring wells and the others are exploration wells. However out of the 425 exploration wells 288 can be used as monitoring wells, 137 are already used as production wells and equipped with pumps and 21 well are abandoned. The newly formed Environment Agency – Abu Dhabi (EAD) is developing a national groundwater monitoring program by consolidating the two projects and developing a central database for water resources in the Emirate. In this paper the well locations were evaluated and optimized using the geostatistical approach. The collected data from both projects over the last 10 years were evaluated and analyzed using the cost effective sampling procedure for reviewing groundwater data and optimizing the national groundwater monitoring network. A data assessment sheet and a lowest-frequency sampling schedule for each groundwater monitoring location were produced. The determination of sampling frequency for a given location is based on trend, variability, and magnitude statistics. The underlying principle is that a location's schedule should be determined primarily by the rate of change in concentrations observed there in the recent past. The larger the rate of change, whether upward or downward, the greater the need for frequent sampling. Applying the cost effective sampling methodology produced, initially, the optimum annual number of required groundwater samples and locations. This will save the annual sampling, data management, and analysis costs.

1 INTRODUCTION

Groundwater monitoring network design, implementation, sampling, evaluation, and optimization is a problem of paramount importance to the environmental field. The collection and analysis of data (chemical, physical, and/or biological) over a sufficient period of time and frequency to determine the status and/or trend in one or more environmental parameters or characteristic directly related to the management objectives for the site in question called monitoring. The environmental observations and data provided from such network are the sole means of assessing and evaluating the statues of groundwater in terms of quality and quantity. These data also could help to develop the future management and protection plans for sustainable development of groundwater resources is arid region where scare water resources are the norm, the availability of water is a key element in determining the scale of long term development. Evaluation and optimization of groundwater monitoring network problem is extremely challenging process because it requires environmentalists/hydrogeologists to capture an impacted system's governing the process, limit monitoring costs, achieve objectives, satisfy the interest of multiple network users (e.g. researchers, public advocators, and decision makers). Also, costs of long term groundwater monitoring represent a significant, persistent, and growing burden for the private entities and government agencies responsible for groundwater management and protection projects, especially as remedies are determined and implemented.

In Abu Dhabi Emirate two monitoring networks for groundwater levels and quality have been developed. The National Drilling Company (NDC) in cooperation with United States Geological Survey (USGS) began development of the first network, begging 1988 and at present continuously monitors about 600 wells in 485 locations and about 48 wells are equipped with automatic monitoring devices (NDC/USGS 2004). In 1996 the German consortium GTZ/DCo under the supervision of Abu Dhabi National Oil Company (ADNOC) began development of the second network which is consists of 675 wells out of them 283 are used as continuous monitoring wells and the others are exploration wells. However out of the 392 exploration wells, 287 are not used as production and can be used as monitoring wells. Figure 1 shows the distribution of the monitoring wells of the two networks (GTZ-DCo/ADNOC 2004, GTZ-DCo/ADNOC 2005). During the development of these two networks there was a lack of communications and there is no cooperation between the two programs. A need exists to evaluate the current two groundwater monitoring networks, develop criteria for future monitoring needs, define the sampling frequency, and combine them into one optimal national monitoring network for the Emirate. The newly formed Environment Agency – Abu Dhabi (EAD) is developing a national groundwater monitoring program by consolidating the two projects and developing a central database for water resources in the Emirate (Dawoud et al. 2005, EAD 2005).

Monitoring networks dimension and sampling frequency reduction (optimization) is a need foreseen because of the already substantial financial effort put into maintaining these networks. The optimization of groundwater monitoring networks plan indicates the specific wells to be sampled and the sampling frequency. Well designed monitoring plan will save time and cost of data

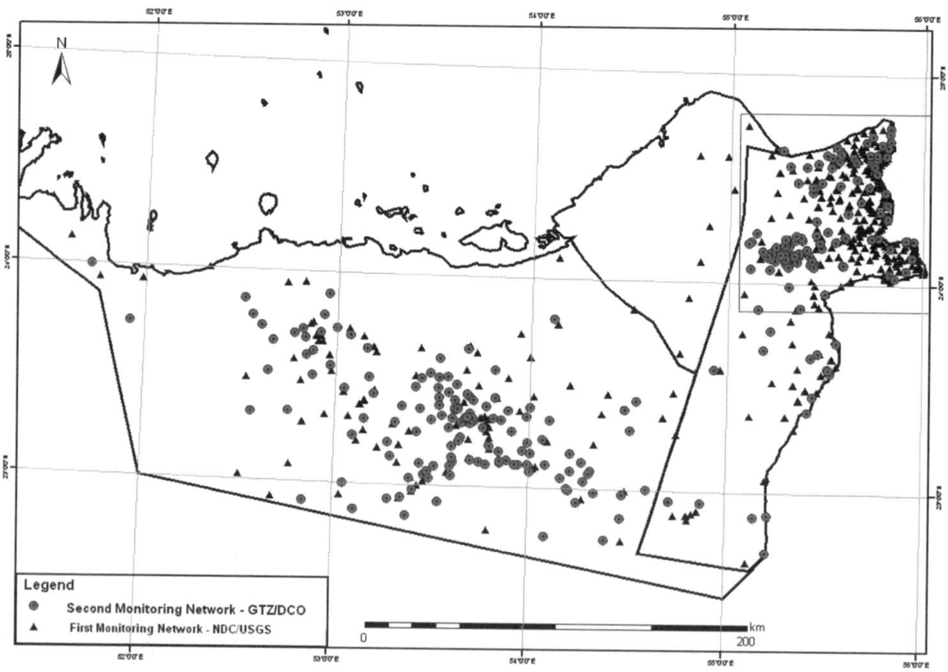

Figure 1. Distribution of the wells of the two networks.

collection, sampling, samples chemical analysis, data storage, data analysis and evaluation. The optimization of long term groundwater monitoring network can be accomplished using variety of approaches. Selecting an approach method involves numerous criteria, the most important of which include the site specific long term performance objectives and the amount and type of available data. Recently, many researchers developed optimization approaches of groundwater monitoring networks such as using statistical and geostatistical methods to identify spatial and temporal redundancies in an existing monitoring network (Stewart 2004, Nunes et al. 2004, Loaiciga 1989). Simulation optimization modeling can be used also to develop a cost effective sampling network design associated with long term monitoring of groundwater (Wu et al. 2005, Reed et al. 2001). The combined simulation and optimization models are appealing because it can account for the complex behavior of groundwater system and identify the best management strategy to achieve a given set of objectives under prescribed constraints (Wagner 1995a, b, Zheng & Wang 2002).

One of the most powerful procedures for reviewing and optimizing the groundwater monitoring network is the Cost-Effective Sampling (CES) program. In 1992 CES program was approved by United State Environmental protection Agency (US EPA) for use (Lamarre et al. 1996). The initial development of the CES program at Lawrence Livermore National laboratory (LLNL) was motivated by an overwhelming number of sampling results with little or no change in concentration over many years (Ridley & MacQueen 2005). This suggested that the monitoring wells were being sampled more often than necessary. CES was developed not only to make sampling frequency recommendations, but also as a data review tool. It provides the researchers and decision makers with a summery of all information required to review and optimize the monitoring network.

In this paper the collected data from both projects over the last 10 years were evaluated and analyzed using the cost effective sampling approach for reviewing groundwater data and optimizing the national groundwater monitoring network for Abu Dhabi Emirate. Data assessment sheet and a lowest-frequency sampling schedule for each groundwater monitoring location were produced. The determination of sampling frequency for a given location is based on trend, variability, and magnitude statistics.

2 CURRENT MONITORING PROGRAMS EVALUATION

Both groundwater monitoring programs in Abu Dhabi Emirate have not, in the past, been implemented on the basis of justification for spatial and temporal intensive monitoring objectives and there was lack in communication and cooperation between the two programs. Cost rationalization and the need to orient one national monitoring program for Abu Dhabi Emirate have made it necessary to optimize and define the gapes in both networks. The new network size, the spatial distribution and sampling frequency must, however, be sufficient to ensure that both spatial and temporal variabilities are correctly included in the optimized national network.

2.1 *First monitoring network (NDC/USGS)*

The first is the joint NDC/USGS Groundwater Research Project which commenced in 1988. About 600 monitoring wells in 485 locations were drilled and monitored regularly. About 48 wells are equipped with automatic loggers and the others are monitored manually twice a year.

2.2 *Second monitoring network (GTZ/DCo)*

The continuous monitoring program has been started in May 1997 where the first monitoring equipment were installed in few wells and increased gradually to cover the whole areas of Abu Dhabi Emirate. At present the monitoring network contains a total of 283 groundwater monitoring wells from both project wells and wells owned by third-parties. 4 monitoring stations were selected in different areas (Ash Shiweib, Al Wijan, Hamim and Liwa) and additionally equipped with 4 meteorological stations. As well as 20 monitoring stations were equipped with rain gauges for measuring

Table 1. Used monitoring devices in the first monitoring network.

Equipment	Description	Measured parameters	Numbers
Floater	Floater Water level Sensor	Water level	75
FloGSM	Floater + GSM	Water level	29
HMG	Hydro-meteorological Station + MDS + GSM	Water level, Electrical conductivity, Temperature, and Meteorological Data	4
Insider	Insider	Water level	93
MDS	Multi Parameter System + GSM	Water level, Electrical conductivity and Temperature	46
MD Soil	Soil Moisture Sensor + MDS	Soil Moisture	2
MPG	Multi Parameter Sensor + Precipitation Gauge + GSM	Water level, Electrical conductivity and Temperature	17
SIN	Shortly Insider	Water level	17

the amount of precipitation. Besides 2 soil moisture sensors were installed into open holes located in Liwa area for measuring the moisture content of the un-saturated zone (vadoze zone). The GSM-Modems were installed into 69 monitoring stations to allow for remote retrieving of the monitoring data.

The EAD's water resources staff visits the monitoring stations every 4 months (3 times annually) for data downloading and routine maintenance of loggers. The monitoring equipment has been designed for continuous recording and storing of groundwater related physical/chemical data in pre-selected time intervals. An in-built microprocessor controls the measuring sensor and stores the data in a memory. All monitoring equipment were manufactured by SEBA Hydrometrie. Different types of groundwater monitoring equipment were used as shown in Table 1. Floater, Insider and Multi Data System (MDS) are the most often used monitoring equipment. Floater and Insider are used for measuring water levels only however MDS is used for measuring the water level, electrical conductivity and groundwater temperature.

The measuring equipment consist of measuring electronics and power supply installed on top of the well, and of sensor, which is lowered about 5 meters below the groundwater level in the well.

Standard procedures have been adapted for the data collection:

- The depth of the water level is measured from the reference point.
- The precise altitude of all reference points have been measured by differential GPS for all monitoring wells.
- The water level sensors are calibrated using water level meters. After calibration the recording of the groundwater measurements can start.

3 OPTIMIZATION PROGRAM

The optimization program includes the clear definition for the objective, document the current monitoring program, examine existing data, check the site candidate for optimization, selection of optimization tools and methodology and the assessment and implementation of the results as shown in Figure 2. The well locations were evaluated and optimized using the geostatistical approach and the temporal data collected from both networks was evaluated and analyzed using the cost effective sampling procedure. Table 2 shows the two used optimization tools, approaches and the required information.

3.1 *Monitoring points spatial distribution and gap definition*

The two monitoring networks will be integrated into one national monitoring network. The geostatistical approach was used to evaluate both monitoring networks spatial distribution. The

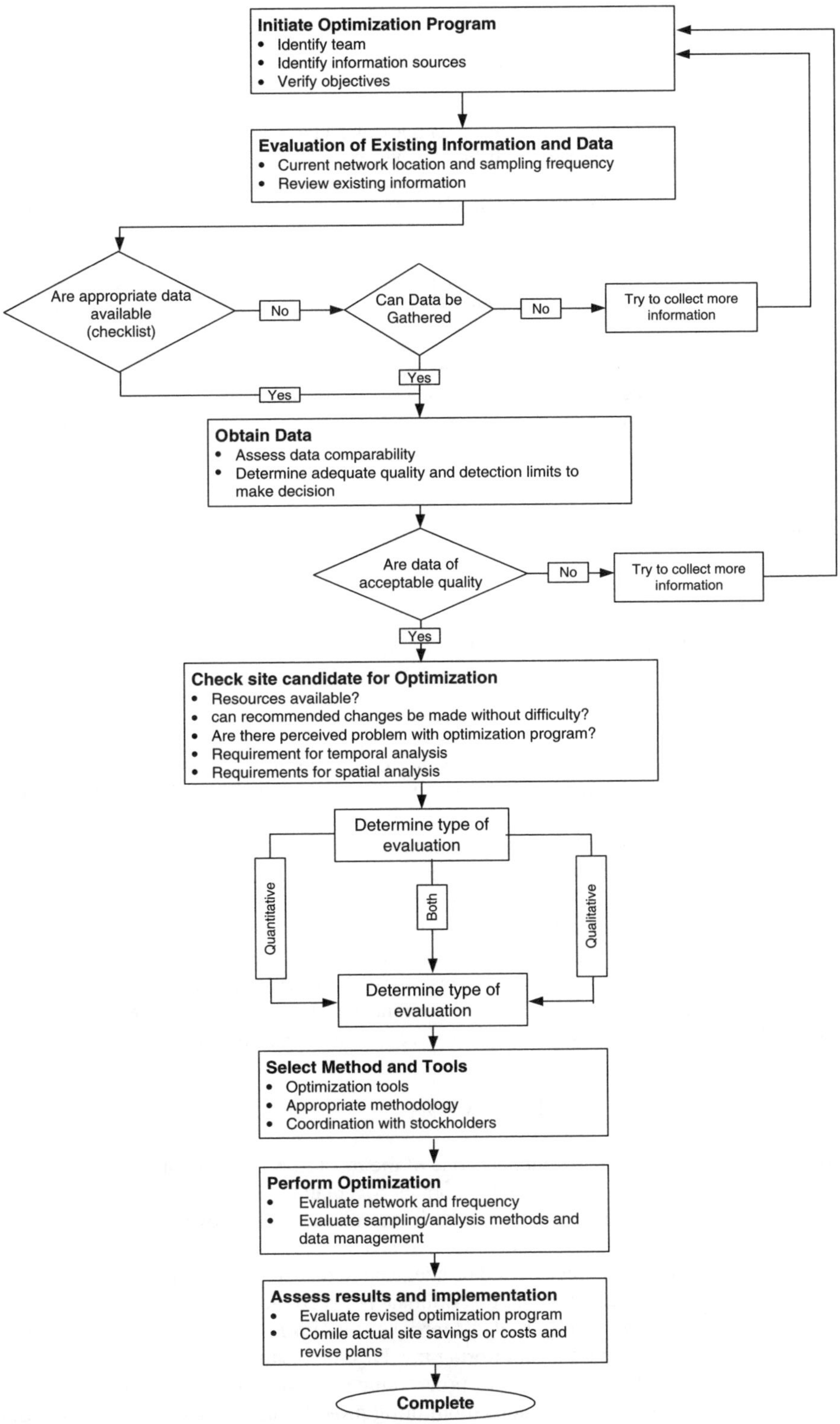

Figure 2. Long term optimization flowchart.

Table 2. Monitoring optimization tools, approaches and data requirements.

Item	Geostatistical spatial optimization (Locations)	Cost effective sampling (Frequency)
Overview	Geostatistical optimization is a spatial algorithm that utilizes geostatistical method to define the network of essential sampling locations. The geostatistical optimization algorithm incorporates a decision pathway analysis that incorporates both spatial and temporal components and is used to identify spatial and temporal redundancies in existing monitoring network.	Cost effective sampling is a methodology for reviewing and assessing the lowest frequency sampling schedule for a given groundwater monitoring location.
Methodology	Weighting scheme utilizing locally weighted quadratic regression examines multiple (time slices) to identify redundant wells based on cost accuracy trade off curves.	Rule-based decision algorithm based on trend, variability, and magnitude statistics recommends optimal frequency at each well.
Data requirements	More than 30 wells (spatial distribution)	• At least 6 monitoring results per well. • Clean down gradient well (guard well).
Appropriate site size	30 to thousands of wells.	Unlimited (well by well analysis) within same operable unit.

parameters that used to define spatial variations in groundwater network are the correlation coefficient and the variogram. This method is commonly used as know as Kriging (Olea 1984, Swain & Sonenshien 1994). This method produces a statistical uncertainty in estimates of unmeasured sites as functions of distance between measured locations by the averages of the values at known points weighted inversely to their uncertainty at the unknown points by minimizing the Kriging variance (Delhomme 1978, Delhomme 1979). An advantage of using the kriging standard deviation as the measure of uncertainty is that its value depends on the semivariogram and the spatial configuration of the observation points, and not on the individual measurements of water level (Olea 1999). A Geographic information system interface was used to allow graphical representations of the results for evaluation. Application of this method indicated that three region needs to fill the gaps as shown in Figure 3. However, two areas in Al Ain and Liwa the monitoring points are very dense.

4 SAMPLING FREQUENCY OPTIMIZATION

Adjusting the frequency of measurements is one of the most important considerations in the optimization of a monitoring program. The development of a plan for water level and quality parameters monitoring that will be used for each well in the observation network is dependent on the objectives of the program and the intended use and level of analysis and accuracy required of the data. The frequency of measurement should be adequate to detect short-term and seasonal groundwater-level fluctuations of interest and to discriminate between the effects of short- and long-term hydrologic stresses as shown in Figure 4. Water-level monitoring may involve "continuous" or periodic measurements. Continuous monitoring involves the installation of automatic water-level sensing and recording instruments that are programmed to make measurements in observation wells at a specified frequency. Continuous monitoring provides the highest level of resolution of water-level fluctuations. Hydrographs constructed from frequent water level measurements collected

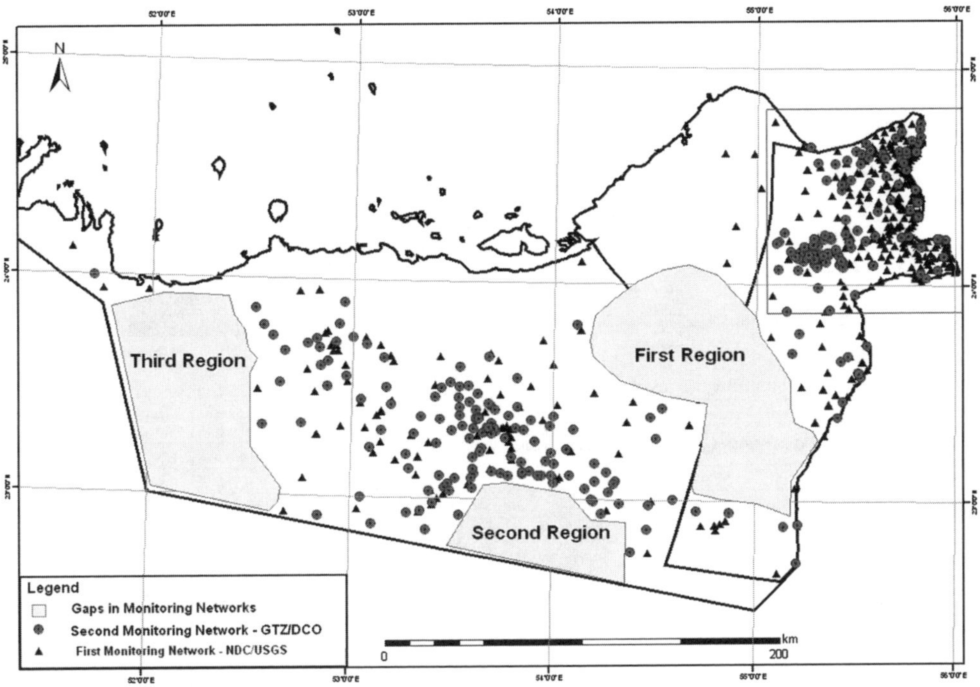

Figure 3. Gaps identification in monitoring network.

with continuous monitoring equipment can be used to accurately identify the effects of various stresses on the aquifer system and to provide the most accurate estimates of maximum and minimum water level fluctuations in aquifers. Figure 5 shows the effect of abstraction from production well on the water level change in nearby observation well. For these reasons, it is often advisable that new observation wells initially be equipped with continuous monitoring equipment to identify stresses on the aquifer and the magnitude and frequency of water level fluctuations. Continuous monitoring may not be required where there is no development nearby the observation well, the hydraulic response of an aquifer to stresses is slow and the frequency and magnitude of water level changes are not great. Near real-time data collection also can be accomplished using a continuous recording device and telecommunication or radio transmitter equipment.

Optimizing the monitoring networks with respect to measurement intensity can be done by minimizing the total cost as a function of the measurement intensity. The total cost as a given measurement intensity could be calculated by summing the cost of measurement and the loss due to limited knowledge as shown in Figure 6. This would be fairly easy if the cost due to limited knowledge (the cost of no monitoring) were known as a function of measurement intensity (Pebesma 1996). Aspects of groundwater monitoring network that control the cost are:

(1) Labours and capital costs of sampling including spatial well density, number of screens sampled per well, measured field parameters, monitoring frequency and measurement intensity.
(2) Chemical analysis cost based on the chemical variables and parameters analyzed.
(3) Network maintenance and operation costs including well rehabilitations, network gaps filling and maintenance of data loggers, pumps and samplers.

The final optimization should take into consideration both the trend in water level changes, the accuracy of required information and the cost as shown in Figure 7. Using this integrated approach indicate that the cost of monitoring can be reduced by about 27% if the sensors (data loggers) time intervals increased to 3 hours instead of 1 hour.

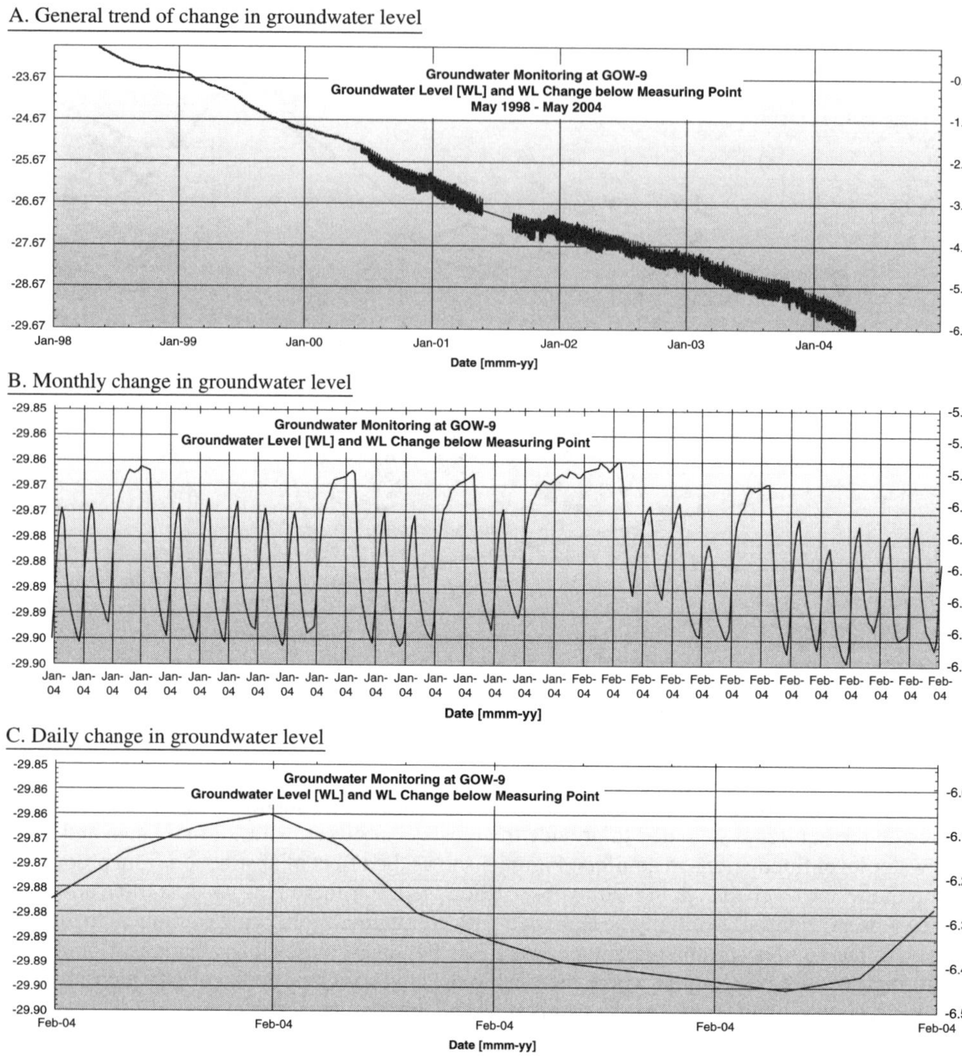

Figure 4. The daily, monthly and annual trends in groundwater levels changes for well GWO-9.

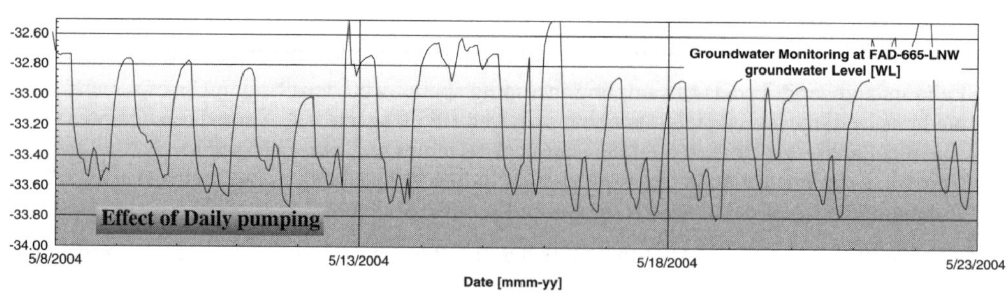

Figure 5. Effect of abstraction for nearby production well on trends in water level changes.

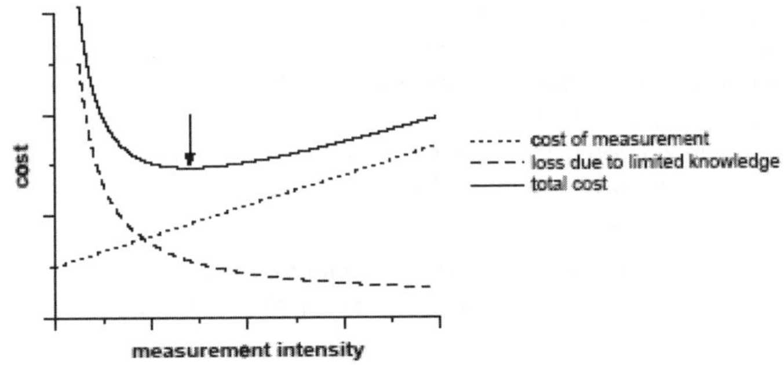

Figure 6. Monitoring network optimization: total cost.

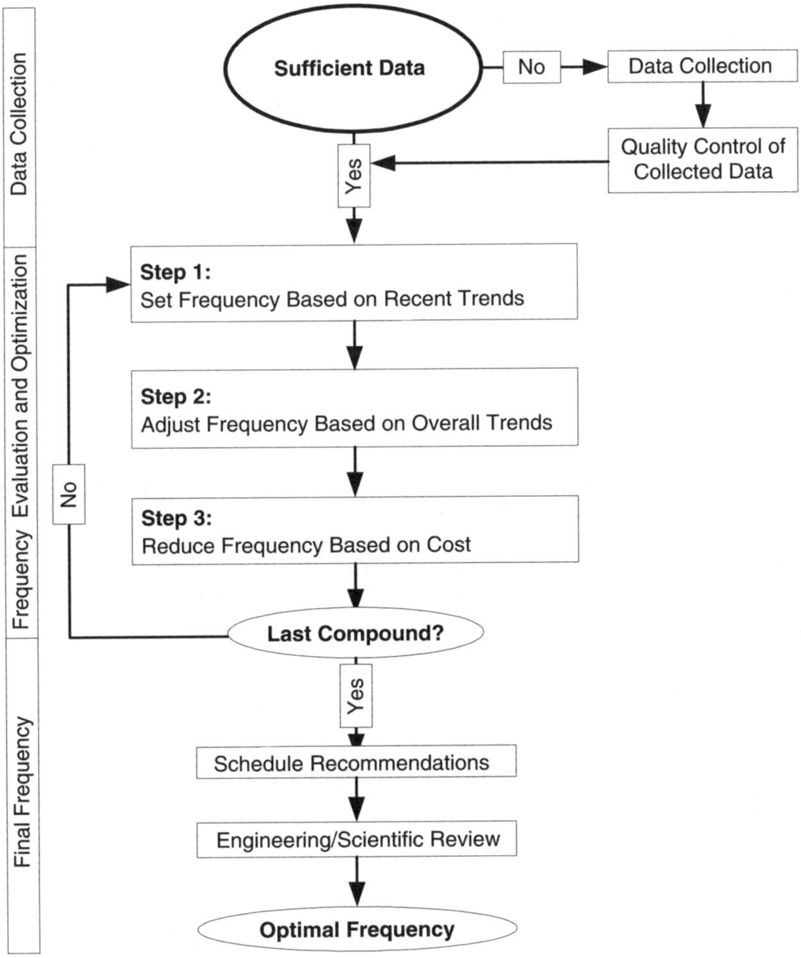

Figure 7. Overview of integrated steps for optimization of sampling frequency.

5 CONCLUSIONS AND RECOMMENDATIONS

The groundwater monitoring networks account for a substantial part of the total cost of the environmental monitoring in the Abu Dhabi emirate. After several years of developing two monitoring network and after nominating EAD to be the responsible agency for water resources management, the question arises whether the current monitoring programs are worth the cost in terms of spatial distribution of monitoring points and sampling frequency. In this study the geostatistical approach was used to evaluate the spatial distribution of the monitoring network and define the gaps, and the cost effective sampling approach was used for adjusting the sampling frequency based on the sampling cost and effect of no sampling locations on the accuracy of data obtained from the network. The following can be recommended:

- Groundwater monitoring networks are very important tool in arid region where the groundwater is vital source for development for evaluating the aquifers potentiality and develop the future management plans.
- Optimization of monitoring networks leads to develop a national monitoring network for Abu Dhabi Emirate and cost reduction.
- The effects of monitoring networks density as shown by comparing results from the optimized national network with the results from the separate two networks are considerable.
- Integration of both groundwater quality and flow together in the optimization process improve the optimization results.
- Measurement errors and absence of quality control on the collected and stored information always result in weak optimization. Reducing measurement errors leads to good optimization and increasing the accuracy of results from the optimized network.

ACKNOWLEDGEMENT

The authors would like to thank HE Majid Al Mansouri the Secretary General of EAD, and Abdulnasser Al Shamsi the Director of Terrestrial Environmental Research Canter, EAD for their kind approval to publish this paper.

REFERENCES

Dawoud, M.A., Farbridge, R., Wonderen, J. van, Brook, M.C. 7 Al Houqani, H. 2005. An ARCGIS database for water supply/demand modeling and management in Abu Dhabi Emirate, UAE. *WASTA Seven Gulf Water Conference Proceedings*: Water in the GCC-Towards integrated Management: 575–592, Kuwait.

Delhomme, J.P. 1978. Kriging in the hydrosciences. *Advances in Water Resources* 1(5): 251–266.

Delhomme, J.P. 1979. Spatial and uncertainty in groundwater flow parameters: a geostatistical approach. *Water Resources Research* 15(2): 269–280.

EAD. 2005. *Development of a Groundwater Monitoring Network for Abu Dhabi Emirate*. Final Report.

GTZ/DCO/ADNOC. 2004. Groundwater Assessment Project Abu Dhabi Status Reports.

GTZ/DCO/ADNOC. 2005. Groundwater Assessment Project Abu Dhabi. Report on the analysis of Core Samples within the GWAP Abu Dhabi 311p.

Lamarre, A.L., Nichols, E.M., Berg, L.L., Dresen, M.D., Gelinas, R.J., Bainer, R.W. & Folsom, E.N. 1996. Compliance Monitoring Plan for the Lawrence Livermore national Laboratory Livermore Site. Report # UCRL-AR-120936, USA.

Loaiciga, H.A. 1989. An optimization approach for groundwater quality monitoring network design. *Water Resources Research* 25(8): 1771–1782.

NDC/USGS. 1997. Ground-Water Research Project for Abu Dhabi Emirate: National Drilling Company and US Geological Survey, Fact Sheet, Abu Dhabi, UAE.

Nunes, L.M., Cunha, M.C. & Ribeiro, L. 2004. Groundwater Monitoring Network Optimization with Redundancy Reduction. ASCE, *Journal of Water Resources Planning and Management* 130(1): 33–43.

Olea, R.A. 1999. *Geostatistics for Engineers and Earth Scientists*. Kluwer Academic Publishers, New York, 303 p.

Olea, R.A. 1984. Sampling design optimization for spatial functions. *Mathematical Geology* 16(4): 369–392.

Pebsema, E.J. 1996. *Mapping groundwater quality in the Netherlands*. Published Ph.D., thesis, Utrecht University, the Netherlands.

Reed, P.M., Minsker, B.S. & Goldberg, D.E. 2001. A multiobjective approach to cost effective long term groundwater monitoring using an elitist nondominated sorted genetic algorithm with historical data. *Journal of Hydroinform* 3(2): 71–89.

Ridely, M. & MacQueen, D. 2005. *Cost Effective Sampling of Groundwater Monitoring Wells: A data Review and well Frequency Evaluation*. Lawrence Livermore National Laboratory technical Report #UCRL-CONF-209770, USA.

Stewart, R.B. 2004. Estimating Cost Savings from the optimization of long-term monitoring programs at US Air Force Bases. *Conference on accelerating site closeout, improving performance and reduction costs through optimization*, Dallas, Texas, USA.

Swain, E.D. & Sonenshien, R.S. 1994. Spatial and temporal statistical analysis of a groundwater level network, Broward County, Florida, U.S. Geological Survey, Water Resources Investigation Report 94-4076.

Wagner, B.J. 1995a. Recent advances in simulation–optimization groundwater management modeling. U.S. national report to international union of geodesy and geophysics 1991–1994. *Review of Geophysics* 1021–1028.

Wagner, B.J. 1995b. Sampling design methods for groundwater modeling under uncertainty. *Water Resources Research* 31(10): 2581–2591.

Wu, J., Zheng, C. & Chien, C.C. 2005. Cost Effective sampling network design for contaminant plume monitoring under general hydrogeological conditions. *Journal of Contaminant Hydrology* 77: 41–65.

Zheng, C. & Wang, P.P. 2002. A field demonstration of the simulation–optimization approach for remediation system design. *Ground Water* 40(3): 258–265.

Arid Land Hydrogeology: In Search of a Solution to a Threatened Resource – Mohamed(ed.)

The dipole flow and reactive tracer test: The coalition forces of science, engineering, and technology attack site characterization drawbacks

M.M. Mohamed[1], S. Banwart[2], D. McKnight[2], N. Thomson[3], S. Thornton[2], R. Wilson[2] & D. Lerner[2]
[1] *United Arab Emirates University, Al-Ain, UAE*
[2] *Groundwater Protection and Restoration Group, University of Sheffield, UK*
[3] *University of Waterloo, Ontario, Canada*

ABSTRACT: Monitored Natural Attenuation (MNA) has become an effective alternative to the more active remediation methods for the in situ treatment of contaminated subsurface environments. This low-cost technology requires a high degree of certainty in the conceptual site model and the underlying reactive transport processes. In order to gather site specific information, site characterization studies must be performed; however, the expense of these studies detracts from the low-cost nature of MNA. Additionally, existing aquifer assessment methods, which range from simple laboratory batch and column studies to large-scale plume studies, suffer from several disadvantages; including the likelihood that the small sample cores are not representative of the heterogeneous subsurface system, the disturbance and possible contamination of cores during collection, the inability to reproduce the exact in situ condition in the laboratory, and the long required time frame. Hence, cost-effective site characterization techniques are required which provide a level of information suitable for demonstrating that in situ processes are present for MNA to be a viable treatment option. Developers of such techniques can not advance without integrating the latest technology in several different research areas such as laboratory methods, field techniques, groundwater mathematical, numerical, and inverse modeling techniques, management and optimization systems, as well as cost-benefit studies.

1 INTRODUCTION

The dipole flow field was originally applied in the late 1980s to remediate contaminated ground water by injecting into one interval and extracting from another interval of a single well. The extracted water was either treated by air stripping within the well, or amendments (e.g., oxygen, nutrients, or chemicals) were added to the extracted water before it was re-injected. Herrling and Stamm (1992) were one of the first to investigate the general features of vertically circulating flow fields in support of there use as a remedial technology.

In an attempt to develop a dependable low-cost site characterization technique, the dipole-flow test, developed by Kabala (1993), is extended to include reactive tracers so that in situ reactive parameters required to support MNA can be estimated. This site characterization test is called The Dipole Flow Reactive Tracer Test (DFRTT). The key concept of the DFRTT is to employ the latest innovations of science, engineering, and technology to characterize the subsurface using the fewest number of field tests. This would lead to major savings in both time and money, which most drawbacks of existing site investigation tools orbit. The DFRTT could be described as a column experiment of an undisturbed uncontaminated large scale field sample.

Extending from the effort of Kabala (1993), Zlotnik and Ledder (1994) developed a number of mathematical models with a focus on understanding the kinematic flow structure around the dipole device, and the drawdown in the well chambers in a uniform anisotropic infinite aquifer. Initially they investigated the region of influence by treating the injection/extraction chambers as point source/sinks and using a Taylor series expansion of the solution of the total drawdown and suggested that drawdown is less than 1% at a radial distance of

$$r > 10\ L\sqrt{K_r / K_z} \tag{1}$$

and a vertical distance of

$$|z| > 10\ L \tag{2}$$

where r is the radial distance from the well, L is half the distance between chamber centers, and K_r and K_z are the radial and vertical hydraulic conductivity. They also considered the drawdown produced by treating the injection/extraction chambers as a linear source/sink rather than as a point source/sink again for an infinite aquifer. A practical outcome from this investigation was equations for the steady-state darcy velocity field as a function of the spatial coordinates (r, z). Using these velocity equations along with the definition of Stokes' stream function for axisymmetric flow, they derived an expression for Stokes' stream function as a function of (r, z). This expression allows for the direct calculation of the pathlines in a dipole flow field (under the stated assumptions).

Zlotnik and Zurbuchen (1998) presented construction details of a dipole system and related components, and provided guidelines for the interpretation of pseudo steady-state field data to estimate hydraulic conductivity. Their interpretation method involves an assumption of steady-state flow and a prior knowledge of the aquifer isotropy ratio ($a^2 = K_r/K_z$). They reported on the results from 153 DFTs conducted in a single well varying the flow rate, and the length of the injection and extraction chambers. In general, the results showed that, for the site investigated, all dipole configurations produced the same hydraulic conductivity profile and compared well with the magnitude and trend produced from grain-size analysis estimates (using the Hazen formula).

Zlotnik et al. (2001) compared hydraulic conductivity estimates from the DFT conducted in 14 boreholes at the Horkheimer Insel field site in Germany to hydraulic conductivity estimated from sieve analysis, constant head permeameter, flow meter, and pump tests. Their conclusions were that the DFT hydraulic conductivity estimates were comparable to the borehole flow-meter results, but overall results from the DFT should be treated as another hydraulic testing technique due in part to the scale issue (spatial scale of $\sim$1 m). Halihan and Zlotnik (2002) presented a modification of the DFT in which the injection and extraction chambers are different lengths and termed this the asymmetric DFT. The testing procedures for an asymmetric DFT are essentially the same as for the symmetric DFT; however, the interpretation method needs to be modified to account for the different chamber lengths. Halihan and Zlotnik (2002) modified the approach used by Zlotnik and Ledder (1994) to produce expressions for the horizontal hydraulic conductivity near the upper and lower chambers. These expressions are functions of the flow rate, steady-state drawdown in each chamber, the chamber length, and shape factor. They used the asymmetric DFT and associated interpretation approach to estimate the horizontal hydraulic conductivity of a fractured dolomite aquifer at the Bissen Quarry site in Wisconsin, USA. The estimated vertical hydraulic conductivity profile from the asymmetric DFTs was similar in order of magnitude to those estimated by the conventional double packer slug extraction tests.

A collective team of researchers from different disciplines including earth scientists, mathematicians, modelers, engineers, chemists, biologists, field managers, as well as laboratory experts are working together in order to develop and enhance such new technology. The objective is to improve the risk based assessment of contaminated land through the development of a reliable low-cost method of site specific characterization of reactive transport properties supported by cost/benefit calculator that will provide decision support to evaluate available site investigation and remediation methods. We term this methodology the in situ Aquifer Assessment Tool (AAT) (Fig. 1). This tool injects water with multiple tracers in one packered section of a borehole, and

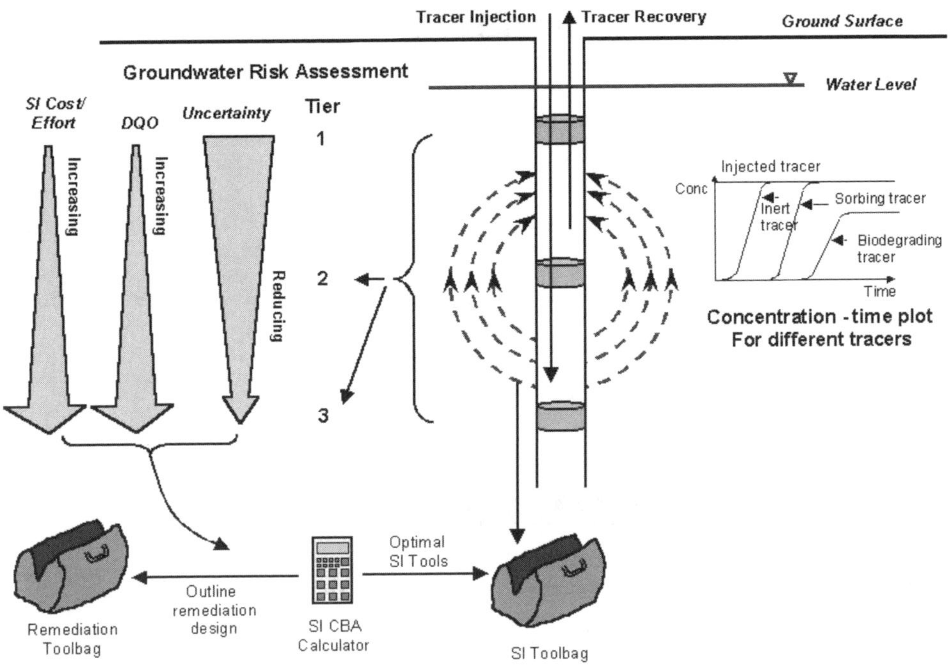

Figure 1. The DFRTT as an in situ Aquifer Assessment Tool.

observes their breakthrough curves in a pumped section of the same borehole. The cost/benefit calculator provides quantitative decision support. It guides when to use the tool and screens potential restoration strategies including MNA.

This research consists of four related major components. Firstly, a DFRTT sandbox-scale physical model was built and will be used to verify theory. Secondly, a numerical multi-phase reactive transport model describing radial flow introduced in DFRTT will be developed. A complete field study including tracer tests at the SIReN site, in great Manchester, UK represents the third component. Finally, a cost/benefit analysis calculator will be developed and tied to the DFRTT for the purpose of assessing the feasibility of MNA.

2 LABORATORY-SCALE EXPERIMENTS

The sandbox physical dimensions are determined by the radial extent of the flow generated by the dipole probe. Hantush (1964) and Zlotnik and Ledder (1994) numerically modelled the dipole flow patterns within a vertical circulation well at steady-state in an unconfined, infinite aquifer and provided the following approximations for the physical extent of a dipole flow field:

$$\begin{aligned}&\text{Minimum horizontal extent of flow} = 10aL \text{ (20aL diametrically)}\\&\text{Minimum vertical extent of flow} = 4L \text{ (8L diametrically)}\end{aligned} \quad (3)$$

where L is the half-chamber separation (or shoulder length; exactly half the distance between chamber centers), and a is the anisotropy ratio ($a^2 = K_r/K_z$, the ratio between the horizontal and vertical hydraulic conductivities). Therefore, a model dipole probe with a shoulder length of 10 cm in an isotropic medium would generate a flow field with a radius of 1 m and would require a sandbox 2 m in diameter in order to minimize boundary interference. Also, the same flow field would extend to $\pm$40 cm from the centre of the dipole probe. A sandbox of dimensions 2 m $\times$ 2 m $\times$ 1 m is constructed (Fig. 2) with the dipole probe (shoulder length not greater than 10 cm) located in a

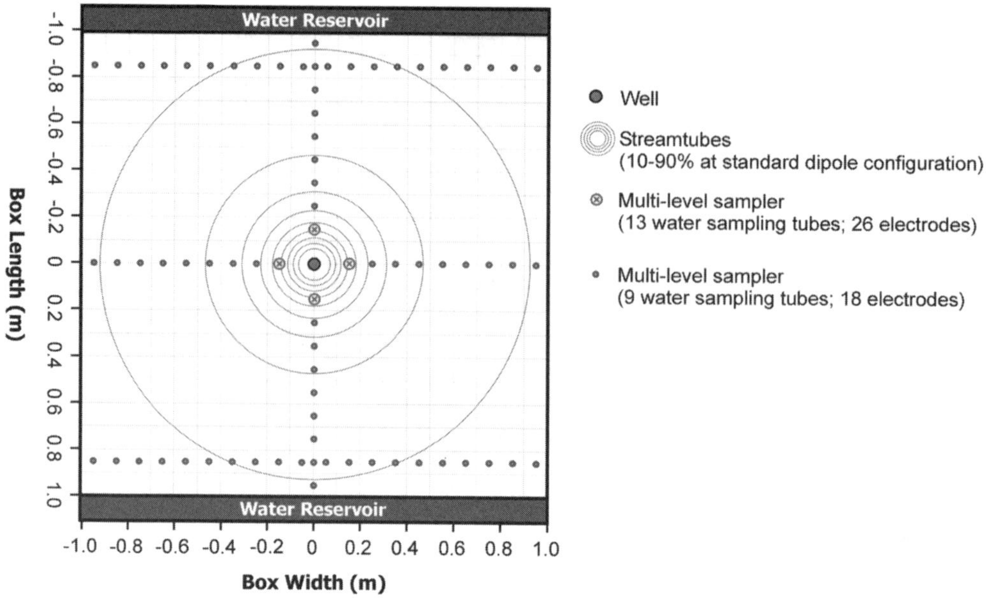

Figure 2. Plan view of the sandbox and the dipole test assembly.

central well. Laboratory-scale trials will not only support the development of the test in the field, but will also provide useful datasets for calibration of the numerical model.

3 FIELD-SCALE EXPERIMENTS

Field trials of the DFRTT will take place at SIReN, a large petro-chemical complex in Greater Manchester, UK. Design of the field experiments was governed largely by the need to recover a suitable amount of tracer in a relatively uncharacterized porous media and also the need to choose a suitable test venue within this large, contaminated site. The field apparatus will consist of a series of three inflatable packers that hydraulically isolate an injection test zone from an extraction test zone and the aquifer above/below the assembly (Fig. 1). The length of the test zones will be determined by the results of the lab-scale model experiments and numerical modeling and practical constraints of construction in the field. However, the extraction test zone will be located above the injection test zone in the assembly. Predictive studies confirm that the field-scale DFRTT will produce breakthrough curves (BTCs) of usable quality provided that significant fractures and layers can be avoided.

4 NUMERICAL MODEL

The DFRTT would not fulfill its potential as an AAT without being accompanied by a numerical model that interprets the specific data required for site investigation and, therefore, increases the popularity of the tool. A multi-phase reactive transport numerical model that can be used to interpret BTCs obtained from a DFRTT is currently under development. The model consists of three major components: a steady-state ground water flow component, a transient aqueous phase reactive transport component, and a reactive solid phase matrix component. This model accounts for well skin effect with a user specified thickness, hydraulic conductivity, and porosity; provides options for the location of upper and lower horizontal boundaries; allows for a user specified location of a horizontal feature with a thickness, hydraulic conductivity, and porosity; and accounts for an asymmetric dipole system. This model is designed to be able to provide an accurate representation of the first-order processes; conform to a variety of field configurations and conditions; computationally handle a range of input parameters; and be extendable so that additional reactions or processes can be added to the model as required with minimal coding effort.

5 COST/BENEFIT ANALYSIS CALCULATOR

The choice of site investigation methods becomes ever more complex with new on-site, in situ (such as AAT) and laboratory based methodologies being developed all the time. In order to facilitate the choice of site investigation techniques, a catalogue of available drilling, sampling and analysis techniques will be constructed including aspects such as relative costs and precision of measurements. Linked to this, a cost-benefit calculator will be constructed in order to return the resolution of measurement required for each parameter at a particular risk assessment tier and ultimately give guidance on the final choice of method with respect to cost.

REFERENCES

Halihan, T., and V. A. Zlotnik. 2002. Asymmetric dipole-flow test in a fractured carbonate aquifer, Ground Water, 40(5), 491–499.

Hantush, M.S., 1964. Hydraulics of wells. Advances in Hydroscience, vol. 1, (edited by V.T. Chow) 281–432, Academic, San Diego, California.

Herrling, B., and J. Stamm. 1992. Numerical results of calculated 3D vertical circulation flows around wells with two screen sections for in situ or on-site aquifer remediation, Computational methods in Water Resources IX.

Kabala, Z. J. 1993. The Dipole flow Test: A new single-borehole test for aquifer characterization, Water Resour. Res., 29(1), 99–107.

Zlotnik, V. A., and G. Ledder. 1994. Effect of boundary conditions on dipole flow, In Computational Methods in Water Resources X, Vol. 2, ed. A. Peters et al., Norwell Mass., Kluwer Acad., 907–914.

Zlotnik, V. A., and B. R. Zurbuchen. 1998 Dipole probe: Design and field applications of a single-borehole device for measurements of vertical variations in hydraulic conductivity, Ground Water, 36(6), 884–893.

Zlotnik, V. A., B. R. Zurbuchen, and T. Prak. 2001. The steady-state dipole-flow test for characterization of hydraulic conductivity statistics in a highly permeable aquifer: Horkheimer Insel Site, Germany, Ground Water, 39(4), 504–516.

Arid Land Hydrogeology: In Search of a Solution to a Threatened Resource – Mohamed(ed.)
© 2006 Taylor & Francis Group, London, ISBN 0 415 41127 0

A study on the water quality and sub-littoral macrobenthos in the vicinity of Madras Atomic Power Station thermal outfall

H.E. Syed Mohamed*, R. Krishnamoorthy[1], V.P. Venugopalan[2] & P.S. Hameed[3]
Environmental Research Laboratory, Dept. of Zoology, Jamal Mohamed College, Tiruchirappalli, India
[1]Department of Biotechnology, Pavender Bharathidasan College of Arts and Science, Tiruchirappalli, India
[2]Water and Steam Chemistry Laboratory, Bhabha Atomic Research Center Facilities, Kalpakkam, India
[3]Director, Environmental Research Center, P.R. Engineering College, Vallam, India

ABSTRACT: Monitoring is necessary to establish the magnitude, spatial distribution and temporal distribution of anthropogenic impact in the receiving environment. The present investigation reports the spatial distribution of temperature, dissolved oxygen and salinity and macro benthos density in Kalpakkam coast where Madras Atomic Power Station (MAPS) discharge the thermal effluent. In the month of March 2002 a boast cruise were undertaken to cover a rectangular area of about 2.7 km^2 and fifteen stations were fixed of which 12 were in the mixing zone (impact area) and remaining 3 were in the control station. The temperature, DO and salinity were measured on the surface and deeper water (2 m to 10 m) at the interval of 1 m. The surface seawater temperature ranged from 29.5°C to 31.60°C. At a depth of 2 m and below the temperature reaches ambient sea temperature levels (29.34°C–29.40°C). The surface DO and salinity concentrations ranged from 5.23 mg/l to 6.47 mg/l and 33.83 ppt to 33.94 ppt respectively. The depth wise measurement of DO concentration decreased with increase in depth (6.14 mg/l–5.43 mg/l) and the salinity concentration increased with increase in depth (33.84 ppt–33.90 ppt). Macro benthos was chiefly represented by Polychaetes, Crustaceans and Molluscs. Among the various stations, the macro benthos density ranged from 144/m^2 to 432/m^2. The macro benthos density at different depths showed variation in the distribution. A comparison of macro benthos density from a control station and impact area of the same depth indicated no significant difference, since the temperature influence is restricted to a maximum of 2 m depth.

1 INTRODUCTION

The gross national product as well as human life expectancy of a nation are directly related to electricity production in the country. An inherent problem in electricity generation is that all the heat energy can not be converted into mechanical energy and a significant part of it has to be rejected in to appropriate heat sinks. Electric power plants are, therefore, often located in the close proximity of natural water bodies such as lakes, rivers, estuaries and oceans to enable them to make use of such water bodies as heat sinks. Electricity production can lead to increase in surrounding temperatures in two distinct ways, namely (a) a direct discharge of heated water, and (b) indirect global warming through CO_2 and other green house gases (GHGs). Rejection of heat to water bodies is unavoidable in electricity generation but needs to be regulated properly so as to avoid undue damage to the ecosystem of the receiving water body. In recent years, the appreciation for such ecological/environmental costs is, slowly but surely increasing all over the world. Sustainable development in the power sector requires use of eco-friendly technologies which aim to keep the productivity high without compromising environmental safety. Apart from the need for preservation of quality of the environment, a holistic understanding of the ecosystem is also necessary to find solutions to a variety of problems that crop up during plant operation. Most of these problems are related either directly or indirectly to biological activity in the water bodies. Therefore, ecological survey of water bodies around power plant is an important endeavor both from the environmental and operational point of view. Hence, the present work was carried out to study the spatial distribution of temperature, dissolved oxygen and salinity and macro benthos density in the vicinity of Madras Atomic Power Station (MAPS).

2 STUDY AREA

Madras Atomic Power Station (MAPS) is located on the south east coast of Bay of Bengal at Kalpakkam (12°33′ N & 80°11′ E), about approximately 65 km south of Madras city (India). The MAPS consisting of two pressurized heavy water reactors (PHWR) of 235 MWe capacity each (presently derated to 170 MWe), Units I and II became operational on 23rd July, 1983 and 18th September, 1985 respectively.

MAPS use coastal seawater at a rate of 35 m^3/sec as a tertiary coolant (ultimate heat sink), employing a once-through type of circuit. The sea water is drawn from about 400 m away from the shoreline, using a 50 m deep buried concrete tunnel, which is a U tube construction. The intake structure is connected to the shore through an approach jetty. Sea water flows by gravity into the fore bay and pump house located on shore, from where it is pumped through the condensers. The heated water is discharged on the shore through an outfall structure (situated on the north of the jetty). The discharged water travels as a canal and mixes with receiving water. The canal formation is due to the sand bar formation along the coast by the interaction of the discharged water flow and the sea current. The direction of flow, the length of the discharge canal and the position of the mixing point (end of the discharge canal where it mixes the sea) vary with the season and are mainly controlled by coastal current and wind direction. After the instantaneous mixing and dispersal of the canal water with the main sea, a thermal plume is formed. The long shore currents carry this plume away.

3 MATERIALS AND METHODS

In the month of March 2002 a boast cruise were undertaken to cover a rectangular area (NE–SW direction) of about 900 m × 3000 m (2.7 km^2) using motorized fishing boat and sampling stations

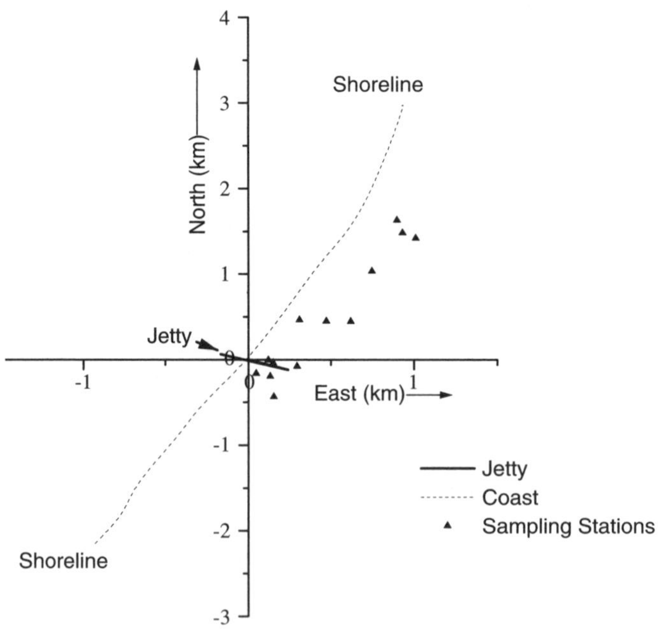

Figure 1. The typical cruise sampling stations.

were marked using the Global Positioning System (GPS). Fifteen stations were fixed of which 12 were in the mixing zone (impact area) which is north of jetty and remaining 3 were in the control station which is south of the jetty where the thermal discharge has no influence on the ambient water (Figure 1). The temperature, DO and salinity were measured on the surface and deeper water (2 m to 10 m) at the interval of 1 m using Multi Probe Water Quality System (Hydrolab, USA, Model DS4a) with an accuracy of 0.1°C. During the cruise, sediment samples were collected from 15 stations using Van Veen grab (area 625 cm^2, penetration 10–15 cm). After the grab was hauled up, the volume of the sediment inside the grab was checked for complete grabbing, and was collected in a plastic bucket. Duplicate samples were collected from each station and average values reckoned in presenting the data. The sediment sample was screened by washing and puddling through a standard circular sieve with a mesh size of 0.5 mm. The sampling technique adopted was that of Holmes and Mc Intyre, 1971.

4 RESULTS AND DISCUSSION

The surface sea water temperature and DO values were plotted as isothermal contours using the software SURFER 6.0 version are given in Figures 2, 4 & 6. In these figures, shoreline is depicted in X-Y coordinates. Intake point is located at the end of the jetty and the jetty itself is indicated as a line perpendicular to the coast. The intersection point of the jetty and the coast is taken as origin (0,0) of the coordinate system. The point marked as 'MP' located on the shoreline indicates the region where the discharge canal empties in to the main sea. The canal *per se* is not shown in figures. The contours have been drawn only encompassing the measured data points. There is no modeling or extrapolation involved in drawing these figures. Distances of all the sampling stations from this point (origin) were calculated from their respective GPS positions.

Temperature data collected in Kalpakkam coast revealed that the surface sea water temperature ranged from 29.5°C to 31.60°C and the ΔT from 0 to 3°C. Figure 2 show changes in the surface temperature pattern of the coastal waters caused by the discharge from MAPS. In front of the mixing point the warm water spreads itself. The benthic area along the shore upto a distance of 200 m

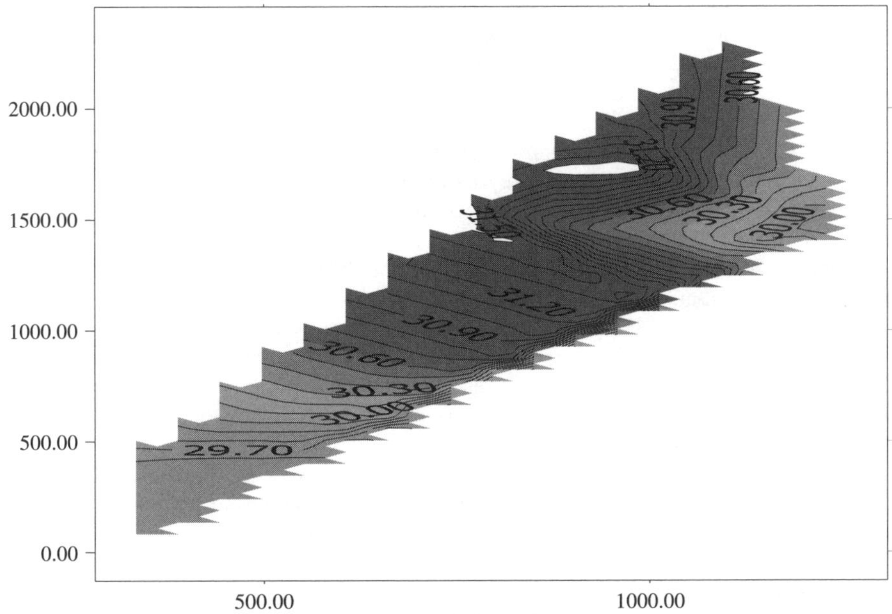

Figure 2. Surface water temperature (°C) in the vicinity of MAPS, Kalpakkam coast during March 2002.

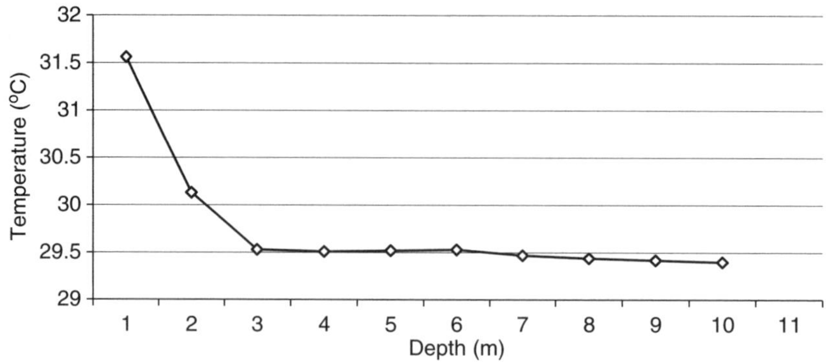

Figure 3. Depthwise temperature (°C) changes in the vicinity of MAPS, Kalpakkam coast during March 2002.

gets exposed to a ΔT of 2–3°C. A ΔT of 1–2°C was observed up to a distance of 900 m North – 400 m south and 400 m into the sea (East) from the mixing zone. The area occupied by the thermal plume was 0.5 sq. km. Beyond 200 m into the sea, the sea bottom is not exposed to warm water due to the buoyant nature of the thermal plume. Hence the total area of the coastal marine environment likely to be impacted due to thermal discharge is a stretch of intertidal zone adjacent to the mixing zone and a stretch of sublittoral zone up to 200 m in the offshore direction.

Depth wise temperature measurement was carried out at the interval of 1 m to a maximum depth of 10 m. From the data it was observed that the temperature decrease was recorded up to 1 m depth and at a depth of 2 m and below the temperature becomes normal. A typical depth wise temperature fluctuation was presented in Figure 3. The surface temperature of 31.39°C decreased to 29.82°C at 1 m depth and from 2 m to 10 m depth the temperature was almost constant (29.34°C–29.40°C). According to Kinne (1967), ocean temperature range decreases with increase in depth. However, within the wave break zone up to the shore, turbulent mixing occurs due to the wave action. Hence, no layering effect is anticipated. The higher surface temperature indicated that the heated water is less denser and has increased buoyancy. Spurr and Scriven (1975) observed the density of the heated water being less than that of the ambient water, heated water plume in the sea is positively buoyant and its direction of movement is determined by coastal wind and current patterns.

The surface sea water dissolved oxygen level fluctuated from 5.23 to 6.47 mg/l and values were plotted as isothermal contours using the software SURFER (Figure 4). Among the different stations studied under the present investigation, an inverse relationship between water temperature and dissolved oxygen was noticed mainly at mixing point. It is observed that stations closer from the mixing point show marginally less DO concentration (5.23–6.17 mg/l) because of the entering of the thermal effluent. A similar observation was also made by Nugent (1970) who has reported that the concentration of dissolved oxygen in the thermal effluent is decreased to about 1 ml/l as it passes through the plant and down the effluent canal in which the temperature noticed was higher than that of ambient water. Heated effluent should predictably cause oxygen depletion in any receiving water (Hynes, 1960). Thames estuary power plant effluents also mildly lowered oxygen concentration though their contribution was relatively minor compared with those of other effluents (DSIR, 1964). In the present study, the marginal reduction of DO could not affect the life processes of the biota especially fishes whose optimal requirement is 5 mg/l (Sverdrup *et al.*, 1962). However, at stations away from the mixing point, there were no significant changes in DO levels as compared to those near the mixing point. Intertidal region the wave action acts to mix atmospheric gases into the water, thus increasing the oxygen content. So wave washed areas never lack for oxygen, because of the constant interaction with the atmosphere and the formation of bubbles plus the stirring up of the substrate.

The depth wise measurement of dissolved oxygen was observed that the dissolved oxygen concentration decreased with increase in depth. In the surface water DO concentration of 6.46 mg/l

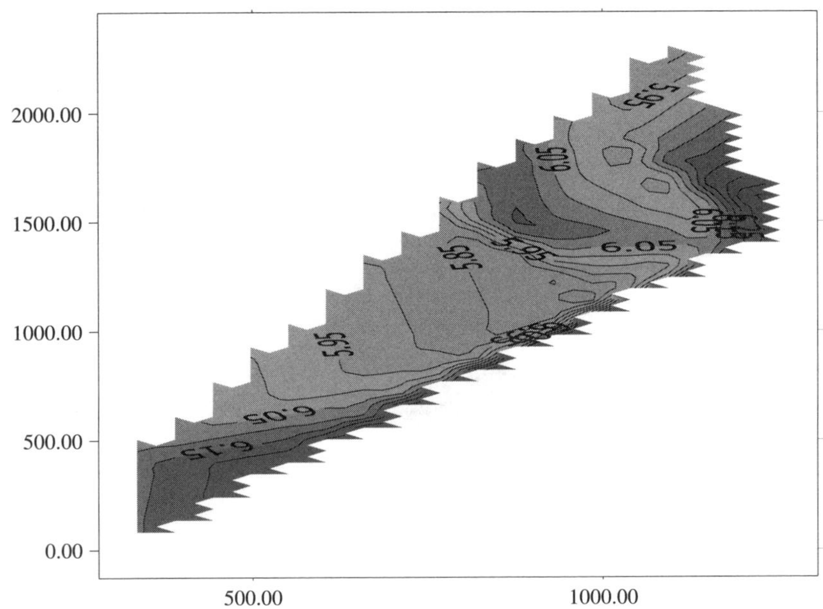

Figure 4. Dissolved oxygen (mg/l) concentration in the vicinity of MAPS, Kalpakkam coast during March 2002.

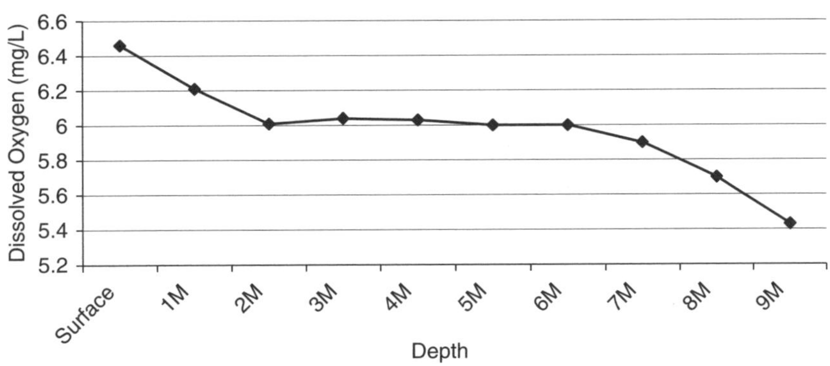

Figure 5. Depthwise dissolved oxygen (mg/l) changes in the vicinity of MAPS, Kalpakkam coast during March 2002.

was recorded and at 10 m depth it was 5.43 mg/l (Figure 5). The low concentration of oxygen in the bottom water was apparently due to the biochemical oxidation of organic matter and its limited renewal. According to Sverdrup *et al.*, (1962) the range of oxygen over the depth of 10 m may be from 0 to 6.4 mg/l.

The surface sea water salinity indicated a minimum value of 33.84 ppt and a maximum value of 33.94 ppt (Figure 6). Among the various stations, difference in salinity level did not show any detectable variation. The higher summer water temperature caused the increased levels of salinity. A similar observation was reported by several authors Balakrishnan Nair *et al.*, 1983; Sankaranarayanan *et al.*, 1984. Rajashree Gouda and Panigrahy, 1992. The reasons for the higher salinity values may be attributed to factors such as evaporation, dominance of neritic water and lack of freshwater in flow during summer months (Murugan and Ayyakkannu 1991; Nair and Ganapathy, 1983). From the depth wise salinity data it was observed that the salinity concentration increased with increase in depth. At 10 m depth salinity reached a level of 33.90 ppt as compared to surface salinity (33.84 ppt).

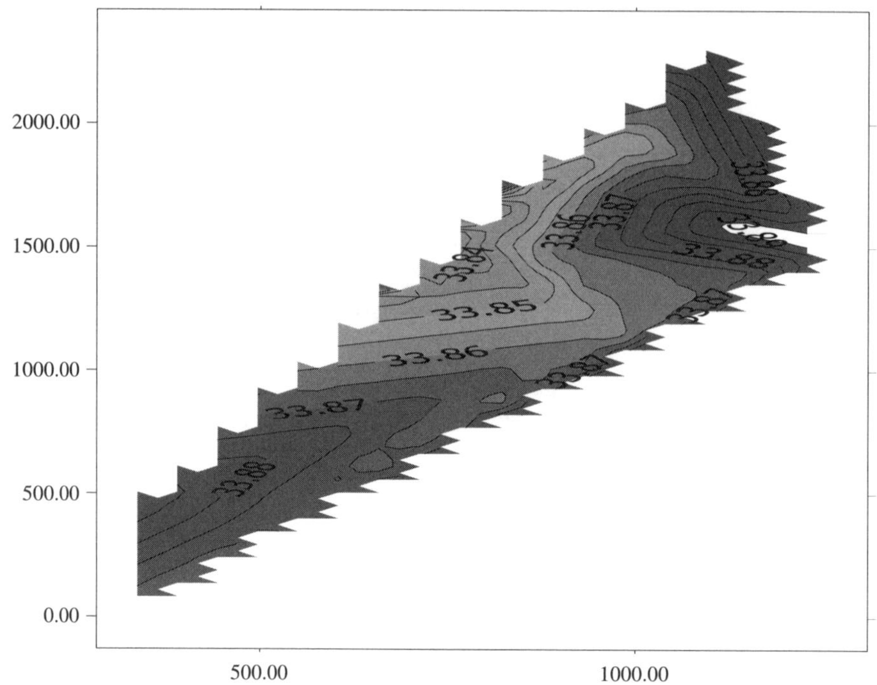

Figure 6. Salinity (ppt) level in the vicinity of MAPS, Kalpakkam coast during March 2002.

Macrobenthos was chiefly constituted by Polychaetes, Crustaceans, Molluscs and miscellaneous species which include fish such as *Therapon spp*. and *Sillago spp*. Polychaetes were the dominant group in the total macrobenthos and they are more or less uniformly distributed in all stations. Among the various stations, the macro benthos density showed the lower level of $144/m^2$ at 3 m depth and the higher level of $432/m^2$ at 9 m depth. The macrobenthos density at different depths showed variations in the distribution. The macrobenthos density recorded at 3 m depth a mean value of $160/m^2$ and at 8 to 10 m depth the density increased to mean value of $416/m^2$. As depth increases, the turbulence of water becomes reduced and in the absence of shifting substratum, as in intertidal zone, the benthos could well settle and flourish at the deeper layers. Also greater availability of organic detritus through sinking of dead and decomposed plants and animals form a rich food sources of these benthic organisms.

A comparison of macrobenthic density from a control station and a station from impact area of the same depth (7 m) indicated no significant difference, since the temperature influence is restricted to a maximum of 2 m depth.

5 CONCLUSION

The present study reported that thermal plume is shore attached it has no bearing on the sublittoral macrobenthic density. The study of distribution of benthos in Kalpakkam coast indicated that the variations in quantity and quality of these animals strongly depend on depth and nature of sediment rather than temperature, since the temperature change in restricted to the surface water of 2 m depth. The dissolved oxygen concentration, one of the most important criteria that sustain life in the aquatic milieu, undergoes only a marginal changes and does not go below the minimum concentration level needed for life sustenance (5 mg/l). Salinity level do not undergo any significant change that is likely to affect marine life. The biological impact of the condenser discharge is limited to a small area in the close vicinity of the mixing point. Large scale damage to the coastal marine ecosystem *per se* or long term impact due to ecosystem level changes attributable to the discharge, are not observed.

ACKNOWLEDGEMENTS

The authors thanks Dr. M. Sheik Mohamed, Principal and Dr. M.M. Shahul Hameed, Head, P.G. Dept. of Zoology, Jamal Mohamed College, Tiruchirappalli for institutional support. The authors are also grateful to Dr. S.V. Narasimhan, Head, WSCL, BARC Facilities, Kalpakkam for technical guidance and Board of Research in Nuclear Sciences, Department of Atomic Energy, Govt. of India for funding the research project.

REFERENCES

Balakrishnan Nair, N., Abdul Azis, P.K., Dharmaraj, K., Arunachalam, M., Krishna Kumar, K. & Balasubramanian, N.K., 1983. Ecology of Indian Estuaries: Part I – Physico-Chemical Features of Water & Sediment Nutrients of Ashtamudi Estuary. *Indian Journal of Marine Sciences*. pp. 143–150.

DSIR, 1964. Effects of polluting discharges on the Thames Estuary. Water Pollution Research Tech. Paper 11, Department of Scientific and Industrial Research, HMSO, London.

Holmes, N.A. & Mc-Intyre, A.D., 1971. *In methods for the study of marine benthos*. I.B.P. Hand book. No. 16 (Blackwell Scientific Publication, Oxford and Edinburgh) pp. 324.

Hynes, H.B.N., 1960. *The biology of polluted water*, Liverpool university press.

Kinne, O., 1967. *Temperature – Invertebrates. In "Marine Ecology"* (O.Kinne, ed.) I(1) pp. 407–514. Wiley, (Interscience), London.

Murugan, A. & Ayyakkannu, K., 1991. Ecology of Uppanar Backwater, Cuddalore: *Mahasagar*. pp. 31–38.

Nair, K.V.K. & Ganapathy, S., 1983. Baseline Ecology of Edaiyur – Sadras Estuarine System at Kalpakkam. *Mahasagar-Bulletin of the National Institute of Oceanograph*, pp. 143–151.

Nugent, R.S., 1970. *The effects of thermal effluent on some macro-fauna of a sub-tropical estuary*. Ph.D. Dissertation, University of Miami, Florida.

Rajashree Gouda & Panigrahy, R.C., 1992. Seasonal distribution and behaviour of silicate in the Rushikulya estuary, East coast of India. *Indian Journal of Marine Sciences*. pp. 111–115.

Sankaranarayanan, V.N., Joseph. T., Jayalakshmy, K.V. & Balachandran, K.K., 1984. A typical Behaviour of Dissolved Silicate in the Cochin Backwater & Periyar River. *Indian Journal of Marine Sciences*. pp. 60–63.

Spurr, G. & Scriven, R.A., 1975. United Kingdom experience of the physical behaviour of heated effluents in the atmosphere and in various types of aquatic system, pp. 227–247 in IAEA (a).

Sverdrup, H.U., Martin W. Johnson & Richard H. Fleming, 1962. *The oceans their physics, chemistry and general biology*, Asia publishing house. pp. 869–673.

Groundwater protection

Arid Land Hydrogeology: In Search of a Solution to a Threatened Resource – Mohamed(ed.)

Role of clay fraction on water content detection in soil using TDR

A.M.O. Mohamed & N. Abbas
Department of Civil and Environmental Engineering, UAE University, Al-Ain, UAE

ABSTRACT: This study seeks to evaluate the role of clay content on water content detection using a TDR probe system and eigendecomposition method. As a demonstration, specimens with variable amounts of bentonite (0 to 18% by wt.) and water contents (0 to 20% by wt.) were compacted at known densities and tested using TDR probe. For each experiment, reflected signals were captured by an oscilloscope and their characteristics were identified using eigendecomposition method. Autoregressive modeling and singular value decomposition methods were utilized for calculating the eigenvalues. The most significant eigenvalues were then identified based on their power. Experimental results indicated that the presented system is sensitive to both water and clay contents in the tested specimens. Multivariate statistical analysis was then performed and the regression analysis resulted in a system of nonlinear equations that best described the system. Actual and predicted results are in agreement, indicating a good performance and versatility of the developed system for wide range of water and clay contents in the tested samples.

1 INTRODUCTION

Time Domain Relectrometry (TDR) detection operates by analyzing the characteristics of a fast rising electric potential signal as it reflects at a probing end of a transmission structure (coaxial cable for example) immersed in a sample under investigation. The characteristics of the reflecting signals are highly influenced by the electrical properties of the sample, which are dependent on the physico-chemical properties of the pore fluid, soil structure and minerals, initial placing conditions, and temperature.

Soil water content, from TDR measurements, is generally calculated using the widely known formula developed by Topp (1980). Topp (1980) found a solitary polynomial function that described the relationship between volumetric water content and apparent dielectric constant for mineral soils, which is given as

$$\theta = -5.3 \cdot 10^{-2} + 2.92 \cdot 10^{-2} \cdot \varepsilon_b - 5.5 \cdot 10^{-4} \cdot {\varepsilon_b}^2 + 4.3 \cdot 10^{-6} \cdot {\varepsilon_b}^3 \tag{1}$$

where θ is the volumetric soil water content, ε_b is the apparent permittivity of the soil. Topp's equation is a pure empirical equation derived from fitting curves of experiment results on a wide range of soils with different moisture contents. However, from the analysis of soil bulk dielectric, we know that volumetric fraction of water is not the only factor contributing to the bulk dielectric constant. Although Topp's equation is regarded as a universal calibration function, immediate concerns should be toward the validity of such equation for predicting moisture content of soils with different textures, surface areas, organic contents, density, and salinity.

In the applications of using TDR system in sandy and loamy soils with little or no electrical conductivity, many researchers reported high accuracy of using Topp's equation. Some of the reported tests are for: (1) fine sand with volumetric water content ranging from 0.05 to 0.35

(Zegelin et al., 1989); (2) sandy loam (Zegelin et al., 1992); (3) sandy and loamy soils with air dry and saturated cases (Topp, 1980); and (4) sand and silty soils (Young et al., 1997).

However, glass beads, clay and high organic content soils showed apparent deviation from this empirical function. The error could be as high as 0.03 in predicted volumetric water content. Some reported tests are: (1) for glass beads with variable diameters ranging from 30 to 450 μm, it was found that the volumetric water content was overestimated (Topp, 1980); (2) for clay soils, it was found that the volumetric water content was underestimated (Topp, 1980); (3) for high organic tropical soils with low density of 0.7 Mg/m^3 and variable volumetric water contents ranging from 0.0 to 0.5, the predicted water content was underestimated (Weitz et al., 1997); (4) soils having high clay and organic contents deviate sharply from Topp's empirical functions.

It is known that soil fractions safeguard our natural waters by attenuating pollutant migration. Soils can adsorb pollutants, buffer pH, aid in the precipitation of pollutants, and host biological activity that may transform or degrade pollutants. Specifically, attenuation occurs by mechanisms such as adsorption, complexation, precipitation, filtration, and biological degradation. The non-living components that may contribute to the retention of pollutants can be either organic or inorganic. The organic fractions are composed of unaltered organic matter such as plant roots, transformed organic matter such as amorphous humic substances and decayed materials (Yong et al., 1992, Mohamed and Antia, 1998). The organic fraction is composed of both crystalline and amorphous materials. The crystalline portion is composed of primary and secondary minerals with primary minerals forming a major portion of sand and silt size particles and a minor part of the clay size fraction. The clay size fraction is composed predominately of layer silicates that are generally described as clay minerals. Mixed in and perhaps coating the soil particles are carbonates and amorphous materials such as oxides, hydroxides, and oxyhydroxides. These components along with the layer lattice clay mineral particles and organic fraction make up the most active components of a soil and control its engineering properties. The question is how much control does each component exercise on the resultant dielectric properties that contribute to the measured response of an electric signal?

Therefore, the overall objectives of the study are: (1) evaluate the contribution of each component in the soil mix design on the response electric signals; (2) utilize the newly developed TDR technique for detecting pollutants in subsurface environment as well as the novel data analysis technique; and (3) develop a method for relating the properties of the response signal to the component mix design.

In this part of the investigation we will be looking to evaluate the effect of clay fraction and water content on the response electric signals and analyze the reflected TDR signals using eigen-decomposition method. The method characterizes the captured signatures by a number of eigenvalues calculated by autoregressive modeling and singular value decomposition methods.

2 THEORY

2.1 *Detection by TDR*

In brief, a monitoring probe, machined from a conducting rod (reference), is connected to a pulse generator and to an oscilloscope via a transmission line (coaxial cables). The oscilloscope is used to acquire the system response through a measurement point on the transmission line. During measurement, an electrical pulse with a fast rising edge is generated periodically by the pulse generator and launched toward the probe. The pulse signal propagating toward the probe appears on the oscilloscope as it passes the measurement point located at a known distance from the generator on the transmission line. As the pulse reaches the probe end, it reflects back on the transmission line with pulse characteristics dependent on the properties of the media surrounding the probe. The reflected pulse signal propagates back toward the generator and, thus, appears on the oscilloscope after a delay time with respect to the time of appearance of the original pulse. The delay time is representative of the distance traveled by the pulse from the measurement point to the probe end and back to the measurement point, as well as the electrical characteristics of the media surrounding the probe. Equally important is the magnitude of the reflected signal, which is also highly dependent on the electrical properties of the composition of the surrounding media.

Captured signals can further be analyzed by different methods such as Fourier spectral analysis (Marple 1987; Kay 1988; Mohamed et al., 2003a), power spectral analysis (Marple 1987; Kay 1988), and eigendecomposition (Press et al., 1992; Richards and Seary 1997; Mohamed et al., 2003b). In this study, eigendecomposition has been utilized because of its ability to isolate secondary signal components containing little power. Such secondary components may not be visible in Fourier spectra.

2.2 *Eigendecomposition*

In eigendecomposition analysis, an orthogonal set of expansion vectors are extracted to indicate dominant variabilities in a collection of data points. The coordinates of the points along each vector will be an eigenvector, and the length of the projection will be an eigenvalue (Gould, 1967, Richards and Seary, 1997). Processing of eigenmodes (eigenvectors and eigenvalues/singular values) of a data series is an important tool in signal analysis (Elsner and Tsonis, 1996). Unlike Fourier decomposition, which partitions signals based on harmonic frequency using parametric sines and cosines, an eigendecomposition partitions signals by their strength using adaptive non-parametric basis functions. Signal components can thus be separated by differences in power. The first eigenmode captures the greatest measures of the variance in the data; the least eigenmode usually captures only the least significant noise oscillations. It does not matter if the component captured is sinusoidal, a square wave, a sawtooth, or anharmonic pattern. Further, the signal may be slowly varying low frequency anharmonic oscillation, or a high frequency sinusoid. The strength of eigendecomposition is the isolation of secondary signal components containing little power. Such secondary components may not be visible in Fourier spectra because of spectral leakage and resolution reasons.

In the eigendecomposition procedure, the main task is to set a sorted eigenvalue threshold whereby the noise that is present in the signal can be removed. All of the eigenmodes are used in the reconstruction of the signal. The principal components of the signal will be captured in the initial eigenmodes while the noise is broken into low power elements. An eigendecomposition can be achieved in any number of ways (Marple, 1987; Kay, 1988; Press et al., 1992; Elsner and Tsonis, 1996). The first step is always the creation of a matrix that uses lagged copies of subsets of the data series. This can be a straightforward data or trajectory matrix, such as the forward prediction (Fwd), backward prediction (Bwd), or forward-backward (FB) prediction matrices used in autoregressive (AR) modeling. These data matrices are usually rectangular, and singular value decomposition (SVD) is used to extract the eigenvectors and singular values. Detailed description of the method of analysis can be found in Mohamed et al. (2003b).

2.3 *Autoregressive modeling (AR)*

In an AR modeling, a value at time t is based upon a linear combination of prior values (Fwd prediction), upon a combination of subsequent values (Bwd prediction), or both (FB prediction). The linear models give rise to rapid robust computations (Marple 1987; Kay 1988). To preserve the degrees of freedom for statistical tests and to furnish a common reference for all AR algorithms, the AR model is represented as follows:

$$y_k = \sum_{j=1}^{M} a_j x_{k+j} \qquad k = M \ldots\ldots\ldots 1 \tag{2}$$

$$y_k = \sum_{j=1}^{M} a_j x_{k-j} \qquad k = M+1 \ldots\ldots\ldots \mathrm{N} \tag{3}$$

In these equations (2 and 3) x is the data series of length N, and a is the autoregressive parameter array of order M. The model is defined as reverse prediction for the first M values, and forward prediction for the remaining N-M values.

The AR coefficients can be computed in a variety of ways. The coefficients can be computed from autocorrelation estimates, from partial autocorrelation (reflection) coefficients, and from least squares matrix procedures. Further, an AR model using the autocorrelation method will depend on the truncation threshold (maximum lag) used to compute the correlations. The partial autocorrelation method will depend on specific definition for the reflection coefficient. The least squares methods will also yield results that are a function of how data are treated at the bounds (matrix size) as well as whether the data matrix or normal equations are fitted.

A basic AR model fit does not offer effective signal noise partitioning (Marple 1987; Kay 1988). Even if pure sinusoids are embedded in a modest level of noise, it may require an order well beyond twice the signal component count to successfully capture and spectrally render the sinusoids. In other words, if the model order is too low, only a portion of the deterministic signal is captured. The reminder is treated as part of the white noise (Gaussian distributed). Spectral components are thus missed. On the other hand, if a model order is too high, the full deterministic signal is captured but some measure of the noise is also modeled. There are three ways to manage this limitation. First, the noise can be filtered or removed prior to analysis. Second, an optimum AR order can be selected that captures all the deterministic signal elements and includes as little noise as possible. The third option consists of an in situ noise removal within the least squares procedures that generate the coefficients, which can be accomplished using singular value decomposition (SVD) in any of the least squares AR models (Marple 1987; Kay 1988).

2.4 *Singular value decomposition (SVD)*

This technique is generally used when Gaussian elimination and Lower and Upper (LU) decomposition fail to give satisfactory results. SVD is also the method of choice for solving most linear least-squares problems. SVD methods are based on the theory of linear algebra as discussed by Press et al. (1992).

The decomposition of any matrix A_{ij} can be expressed as a sum of outer products of columns of U and rows of VT, with the "weighting factors" being the singular values w_j.

$$A_{ij} = \sum_{k=1}^{N} w_k U_{ik} V_{jk} \tag{4}$$

where U and V are orthogonal matrices in the sense that their columns are orthonormal.

2.5 *Solution by use of singular value decomposition (SVD)*

Consider the set of simultaneous Eqs. 1 and 2 and rewrite them in a general form as:

$$y = A \bullet x \tag{5}$$

where y is the predicted vector, $\boldsymbol{A}$ is the autoregressive parameter array and x is a vector containing the data series.

The least-squares solution vector x is given by:

$$x = V \bullet \left[diag(1/w_j) \right] \bullet (U^T \bullet y) \tag{6}$$

To obtain a solution, one finds x that minimizes

$$\chi^2 = \left| A \bullet x - y \right|^2 \tag{7}$$

Then, the solution of Eq. 7, can be written as:

$$x = \sum_{i=1}^{M} \left(\frac{U_i \bullet y}{w_i} \right) V_i \tag{8}$$

The variance in the estimate of a parameter x_j is given by:

$$\sigma^2(x_j) = \sum_{i=1}^{M} \frac{1}{w_i^2} (V_i)_j^2 = \sum_{i=1}^{M} \left(\frac{V_{ji}}{w_i} \right)^2 \tag{9}$$

and the covariance is given by:

$$Cov(x_j, x_k) = \sum_{i=1}^{M} \left(\frac{V_{ij} V_{ki}}{w_i^2} \right) \tag{10}$$

3 EXPERIMENTATION

3.1 *TDR system setup*

The schematic diagram shown in Figure 1 illustrates the arrangement of the measurement system. The pulse generator is connected via a transmission line (coaxial cable) to the output (OUT) port of the directional coupler and the probe is connected via another transmission line to the input (IN) port of the directional coupler. An oscilloscope is connected to the coupled (CPL) port of the directional coupler via a third transmission line. In this arrangement, the output signal from the pulse generator is transmitted only to the probe while the reflected signal is coupled to the oscilloscope for recording. Since, incident pulses do not appear on the oscilloscope, signal separations into incident and reflected become an easy task for further data analysis.

The probe consists of a signal line and a ground conductor and terminates at the coaxial cable end at the sample side.The probe structure was fabricated using a 90 mm long aluminum rod with a diameter of 10 mm and a ground disc. The probe length was determined by considering the probe as a monopole structure, which has an infinite ground plane (Said et al., 2001; Mohamed et al., 2001; 2002; 2003a&b). If such a plane is folded in cylindrical coordinate system, then a coaxial transmission line is obtained. In the designed probe, the ground plate (the circular disk) has a relatively small extent. To extend the ground plane of the probe, while maintaining the compactness of the probe design, two conducting rods have been attached to the ground plane. Adding the two conducting rods to the ground plane does not alter the probe structure to a three-wire transmission line. In fact adding more conducting rods to the ground plane will enhance the coupling between the monopole and the ground plane, but would on the other hand limit the extent of the area being

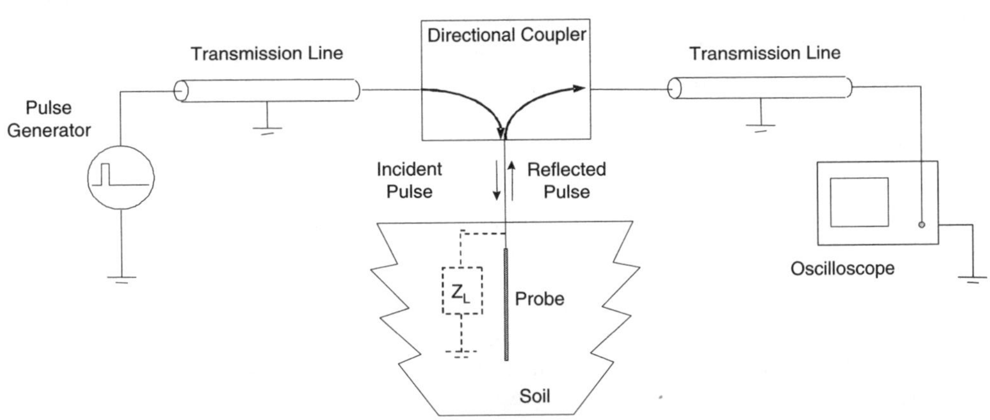

Figure 1. Schematic drawing of the used TDR system arrangement showing the pulse generator, probe, oscilloscope, and the directional coupler which isolates the reflected signal for simplified signal processing.

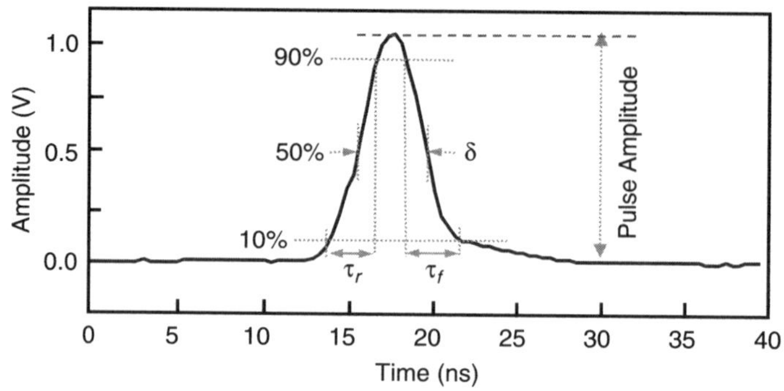

Figure 2. A plot of the generated pulse used in the measuring system as captured on the oscilloscope.

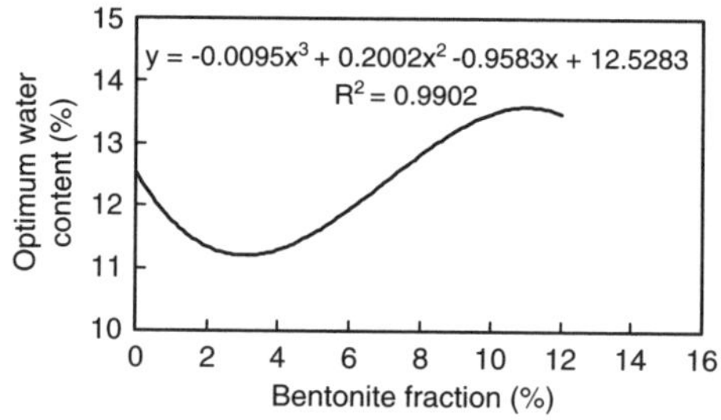

Figure 3. Maximum dry density variations with percent bentonite fraction in the sample.

probed and add difficulty to inserting the probe in solid media, such as sand. The attached rods had a diameter measuring 2.5 mm and were placed at a distance of 17.5 mm, center-to-center, from the monitoring probe.

The used coaxial cables had a line capacitance of 101 pF/m and an impedance of 50 Ω for connecting the pulse generator, the measuring probe and the oscilloscope. A 3.6 m long cable was used to allow enough time delay for the reflected pulse to appear on the oscilloscope screen. The pulse repetition rate was 8.3 MHz, amplitude of 4 V, and rise and fall time was about 3 nano-seconds. Figure 2 shows a plot of the applied pulse as captured by the oscilloscope with an illustration of the different pulse parameters. It should be pointed out that each pulse is composed of infinite number of sinusoidal harmonic frequencies of the fundamental frequency 8.3 MHz. These harmonics are the Fourier components of the pulse.

3.2 *Materials*

The soil material used in this investigation is sand dune obtained from Al Ain district, UAE, mixed with different amounts of bentonite obtained from Abu Dhabi National Oil Company (ADNOC), Abu Dhabi, UAE. The selected testing procedures were carried out following procedures described by ASTM standards (ASTM 1992).

The experimental results indicated that the sand dunes had variable diameters ranging from 0.1 to 1 mm, specific gravity of 2.53, maximum dry density of 1.53 Mg/m^3, and optimum moisture content of 10%. The compaction characteristics of the mixed soil samples are shown in Figures 3 and 4. It can be seen that as the bentonite fraction increases the maximum dry density increases due to the increase of bentonite coatings to the sand grains and decreasing the voids ratio. At low added bentonite fraction (0–4%) the optimum water content decrease because the available water was used to hydrate the bentonite and some of the water used for hydration, especially that in Stern layer, was

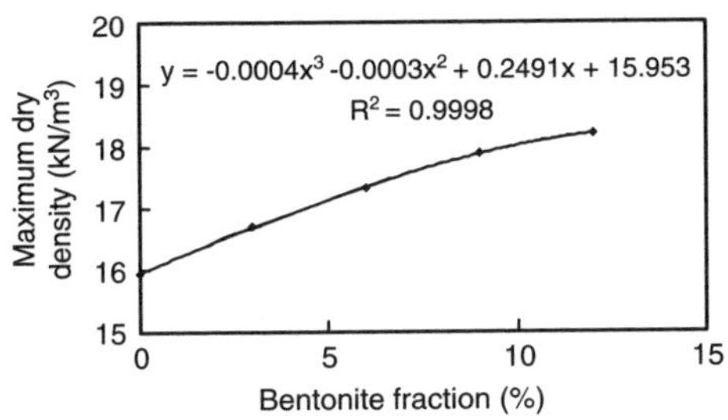

Figure 4. Optimum water content variations with percent bentonite fraction in the sample.

not easily removed during the water content determination by the standard procedures. Beyond the bentonite fraction of 4%, the optimum water content increases as expected.

3.3 *Methods*

Various sets of samples were prepared to study the system response to moisture and bentonite fraction contents. The bentonite fraction varied from 0% to 18% by weight with increments of 3% while the moisture content varied from 0 to 20% by weight in increments of 4%. Deionized water was used in all experiments. To ensure homogeneity of moisture distribution, samples were prepared about 8–10 hours in advance of testing. Soil material was then placed in a testing column having 60 mm in diameter and 120 mm in length at a constant dry density of 1.62 Mg/m^3. It is worth noting that this low density was chosen because of the difficulties we have encountered during probe installation. It is understandable because of the high swelling capacity of bentonite, which made it difficult to drive the probe into the soil. For probe installation, a hole with a diameter slightly smaller less than that of the probe size was drilled through the compacted specimen and the testing probe was pushed through the compacted soil material. Then, the probe was connected to the oscilloscope as shown in Figure 1.

The setup of the system was adjusted, so that the oscilloscope could take 32 readings at a time and calculate the average. Special computer software (FlukeViewTM) was used to acquire the data measured for each test. Reflected signals were captured and recorded for each variation of moisture content and/or bentonite fraction.

4 RESULTS AND DISCUSSION

4.1 *Time domain results*

Figure 5 shows the captured waveform for the probe held in air while Figure 6 shows the captured waveforms for soil mixed with deionized water at various water contents by weight and 9% bentonite fraction. The results indicate that as the moisture content increases, signal peak broadening as well as the signal width at 50% of its peak value increases. Also, as the moisture content increases, signal amplitude decreases and time delay increases. Normalizing the results shown in Figure 6 with respect to that shown in Figure 4, one can easily observe that as the moisture content increases, the relative signal amplitude decreases. This could be attributed to increase of medium capacitance and decrease of its resistance (Mohamed et al., 2001& 2002).

To demonstrate the effect of bentonite fraction, reflected signals are presented for bentonite fractions of 4, 8, and 12% and water content of 3% as shown in Figures 7. The results in Figure 7 clearly indicate that as bentonite fraction increases, signal width at 50% of its peak decreases. Furthermore, as bentonite fraction increases, reflected signal amplitude decreases and the time delay increases. In addition, as bentonite fraction increases, the relative amplitude also decreases. This is so because as bentonite fraction increases the soil medium dielectric constant and capacitance

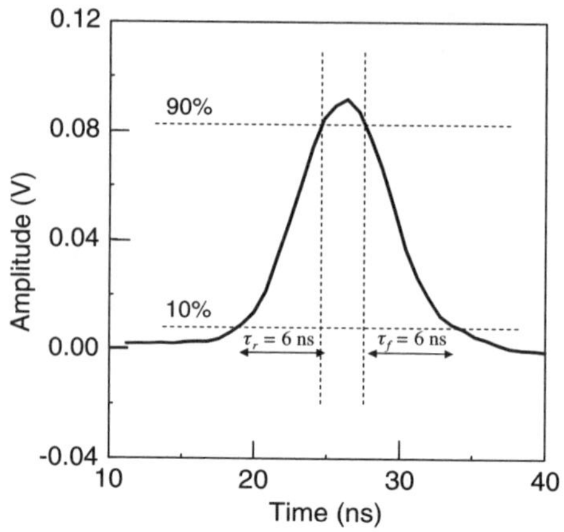

Figure 5. TDR reflected signal as captured on the oscilloscope, while the probe is held in air.

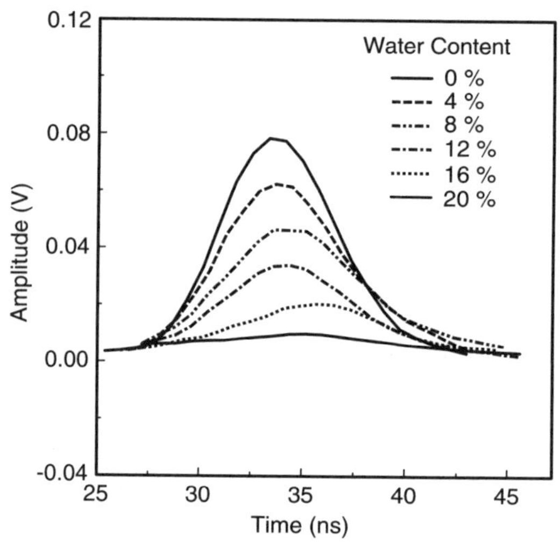

Figure 6. TDR reflected signal as captured on the oscilloscope for various pore deionized water content and 9% bentonite fraction.

increases and its resistance decreases. These results are consistent with diffuse ion theory in which as dielectric constant increases soil water potential increases and medium capacitance increases (Mohamed and Antia, 1998). To further analyze the preceding time domain experimental data, one uses the eigendecomposition technique that described above.

4.2 *Eigendecomposition analysis*

The eigendecomposition technique was utilized to analyze the experimental results for soils tested with various water and bentonite fractions. Typical results for the calculated eigenvalues are shown in Figures 8 and 9 for bentonite fractions of 3 and 12%, respectively. These figures show the variations of the calculated eigenvalues (E_1 to E_5) with water and bentonite fractions.

The results indicate that there are distinct variations at the first and second eigenvalues that represent the largest variations of the data as well the maximum power in the signal. For the same order of the eigenvalue and bentonite fraction, as water content increases the magnitude of the

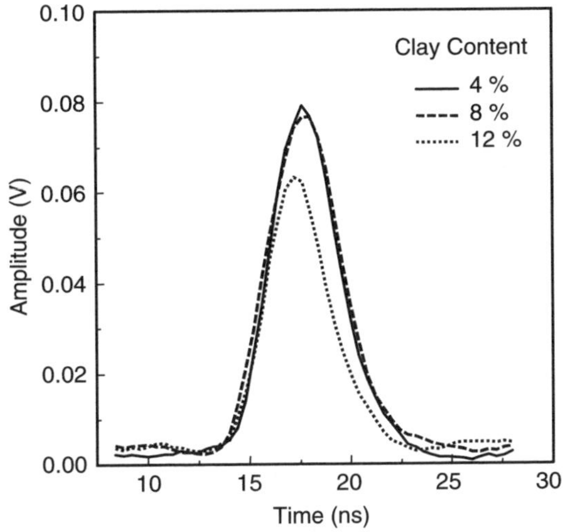

Figure 7. TDR reflected signal as captured on the oscilloscope for various bentonite fraction and water content of 3%.

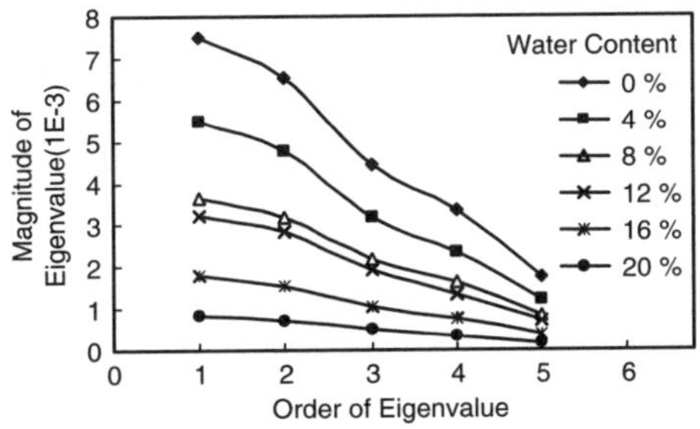

Figure 8. Variations of first five extracted eigenvalues as a function of water contents for soil samples with bentonite fraction of 3%.

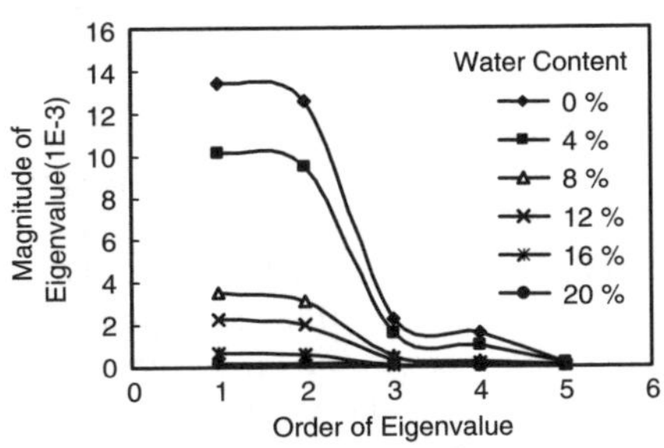

Figure 9. Variations of first five extracted eigenvalues as a function of water contents for soil samples with bentonite fraction of 12%.

eigenvalues decreased (Figures 8 and 9). However, as the bentonite fraction increases, the difference in the magnitude of the first and second eigenvalues becomes larger than that of the remaining eigenvalues. This can be advantageously utilized to detect presence of clay content in subsurface soils. In addition, results shown in Figures 8 and 9 indicate that the variations of E_1 and E_2 are noticeable while E_3 to E_5 are not.

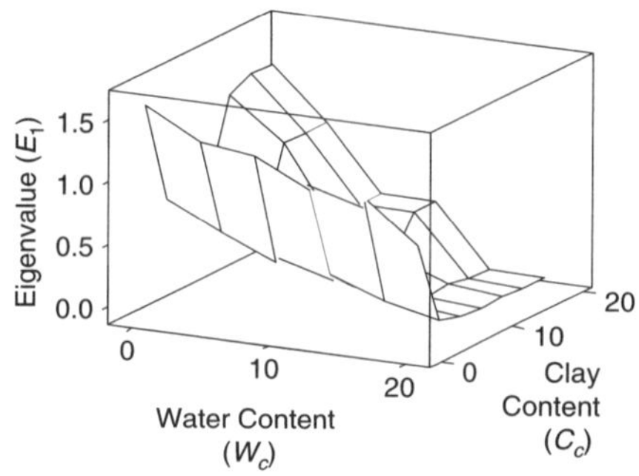

Figure 10(a). A three dimensional representation of the variations of the first eigenvalue with water content (W_c) and clay content (C_c).

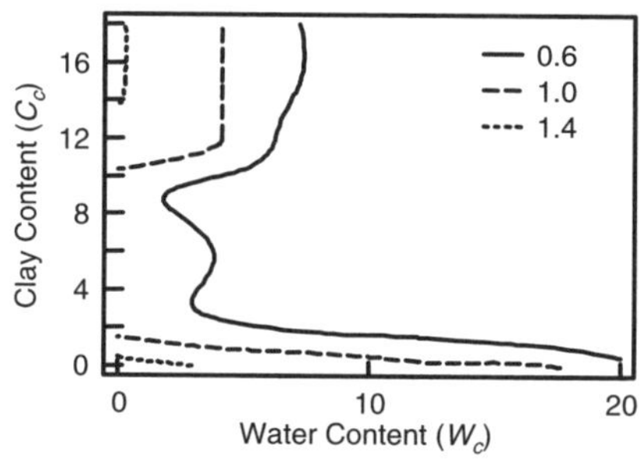

Figure 10(b). A contour representation of the variations of the first eigenvalue with water content (W_c) and clay content (C_c).

4.3 *Determination of water content and clay fraction*

The results recorded from TDR and the first and second eignvalues obtained from eigendecomposition were analyzed using MINITAB to obtain relationships between the first two orders of eignvalues and water content and bentonite fraction (clay content). A three dimensional plot of the variations of the first order eignvalue with respect to water content and clay content is shown in Figure 10a, while Figure 10b shows the contour plot of the first eigenvalue, water content and clay content.

The following equations represent the nonlinear regression of the results:

$$E_1 = 1.37 - 0.0469 \times W_c - 0.122 \times C_c + 0.00068 \times W_c^2 - 0.00772 \times C_c^2 \\ - 0.00384 \times W_c \times C_c - 0.000081 \times W_c \times C_c^2 + 0.000142 \times C_c \times W_c^2 \quad (11)$$

where E_1 is the first eignvalue of the reflected signal; W_C is the water content as a percent; C_C is the clay content as a percent. The squared residuals of the fitting equation are 85.1%. That means the fitted regression model explains about 85.1% of the observed variations in amplitude [R^2 = 85.1%; R^2 (adjusted) = 82.0%]. The observed significance levels (*p*-values) of the T-test are all $\leqslant$0.639. Furthermore, the analysis of variance (ANOVA) F-test found to be 27.7, with 7 and 34 degrees of freedom, which indicates that the regression model is highly significant overall (the corresponding p-value is less than 0.0001). Also, the coefficient of variation is found to be small (*s*/mean = 0.2073/0.5505 = 37.7%). The predicted three dimensional representations of the variations of the first eignvalue E_1 with water content W_C and clay content C_C are shown in Figure 10c.

Similarly, the following nonlinear equation is obtained for E_2:

$$E_2 = 1.04 - 0.0339 \times W_c - 0.0753 \times C_c + 0.00042 \times W_c^2 - 0.00605 \times C_c^2 \\ - 0.00497 \times W_c \times C_c - 0.000075 \times W_c \times C_c^2 + 0.000171 \times C_c \times W_c^2 \quad (12)$$

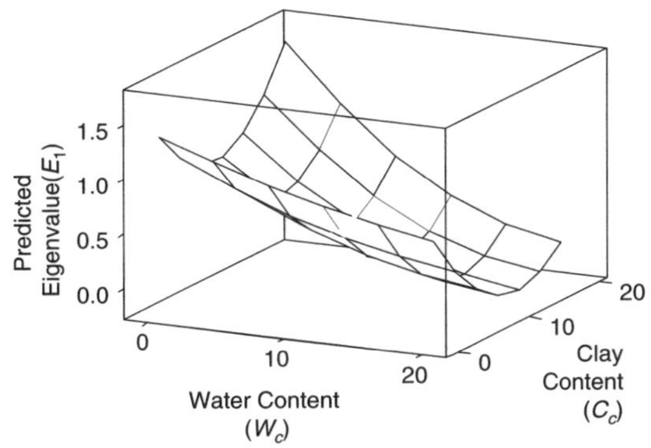

Figure 10(c). Predicted three dimensional representation of the variations of the first eigenvalue with water content (W_c) and clay content (C_c).

where E_2 is the second eignevalue. The observed significance level (p-values) of the T-test are all ≤ 0.709. Furthermore, the ANOVA F-test is found to be 37.14, with 7 and 34 degrees of freedom, which indicates that the regression model is highly significant overall (the corresponding p-value is less than 0.0001). In fact the fitted regression explains about 88.4% of the observed variations [$R^2 = 88.4\%$; R^2 (adjusted) $= 88.4\%$]. Also the coefficient of variation is found to be small (s/mean = 0.1602/0.4812 = 33.3%).

5 SUMMARY AND CONCLUSION

In this study, water contents in soils containing different fractions of bentonite are detected using a TDR probe system and eigendecomposition technique. Samples were prepared with different combinations of deionized water and bentonite fractions. For each experiment, reflected signals were captured by an oscilloscope and their characteristics were identified using eigendecomposition technique. Autoregressive modeling and singular value decomposition methods were utilized for calculating the eigenvalues. The most significant eigenvalues were then identified based on their power. The results indicated that the TDR probe is capable of detecting water content and clay content in tested specimens. Multivariate statistical analysis was performed and a system of nonlinear equations was developed. Actual and predicted results are in agreement, indicating a good performance and versatility of the developed system for wide range of formation water content and clay fraction content in soils. Therefore, with the use of TDR probe and the eigendecomposition technique one could predict the water content and clay content in subsurface soils.

ACKNOWLEDGEMENT

This study was supported by a research grant awarded by the Research Affairs at the UAE University.

REFERENCES

American Society for Testing and Materials (ASTM) (1992) Annual Book of ASTM Standards, ASTM, Philadelphia.

Dowding CH (ed.) (2001) Proceedings of the Second International Symposium and Workshop on Time Domain Reflectometry for Innovative Geotechnical Applications, September 5–7, 2001, Evanston, Illinois, ISBN 0-9712631-0-8.

Elrick EE, Kachanoski RG, Pringle EA, Ward A (1992) Parameter estimation of field scale transport models based on time domain reflectometry measurements, Soil Sci. Soc. Am. J., 56:1663–1666.

Elsner JB, Tsonis AA (1996) Singular spectrum analysis, Plenum Press.

Gould P (1967) The geographical interpretation of eigenvalues, Institute of British Geographers Transactions, 42:53–85.

Heimovaara TJ, Bouten W (1990) A computer-controlled 36- channel time domain reflectometry system for monitoring soil water contents Water Resources Research, 26:2311–2316.

Kachanoski RG, Pringle E, Ward A (1992) Field measurement of solute travel times using time domain reflectometry, Soil Sci. Soc. Am. J., 56:47–52.

Kay SM (1988) Modern spectral estimation, Prentice-Hall, pp. 153–270.

Mallants D, Vanclooster M, Meddahi M, Feyen J (1994a) Estimating solute transport parameters on undisturbed soil columns using time domain reflectometry, J. Contaminant Hydrology, 17:91–109.

Mallants D, Toride N, Vanclooster M, van Genuchten M Th, Feyen J (1994b) Using TDR to monitor solute transport in long undisturbed soil columns during steady saturated flow, In: Proceedings of the 15th International Congress of Soil Science, Acapulco, Mexico, July 10–16, 147–148.

Marple SL Jr (1987) Digital spectral analysis with applications, Prentice-Hall, pp. 172–284.

Mohamed AMO (2002) Time domain reflectometry: operation principles and its potential application for determination of subsurface concentrations of pollutants, In: Developments in Arid Regions Research, 2, A.A. Balkema, Alawaji (ed.), 315–326.

Mohamed AMO, Antia HE (1998) Geoenvironmental engineering, Elsevier, Amsterdam, 707p.

Mohamed AMO, Said RA, Al-Shawawreh NK (2000) Development of an electrical polarization technique for subsurface pollutants detection, In: Geoengineering in Arid Lands, Mohamed & Al Hosani (eds) ©2000 Balkema, Rotterdam, ISBN 90 5809 160 0, 633–639.

Mohamed AMO, Said RA, Al-Shawawreh NK (2001) A TDR system for subsurface pollutants detection (II): application and analysis, Second Int. Symp. and Workshop on Time Domain Reflectometry for Innovative Geotechnical Applications, Dowding, C.H. (ed.), ©2001 by Infrastructure Technology Institute, Evanston, IL 60201, ISBN 0-9712631-0-8, 371–380.

Mohamed AMO, Said RA, Al-Shawawreh NK (2002) Development of a methodology for evaluating subsurface concentrations of pollutants using electrical polarization technique, Geotechnical Testing Journal, GTJODJ, 25(2):157–167.

Mohamed AMO, Said RA, Al-Shawawreh NK, El-Bassiouni MY (2003a) Evaluation of water content and ionic concentrations of soils via frequency domain analysis of TDR waveforms, International Journal of Subsurface Sensing Technologies and Applications, 4(3):159–186.

Mohamed AMO, Said RA, El-Bassiouni MY (2003b) Eigendecomposition of TDR waveforms: A novel method to determine water content and pore fluid concentration of sandy soils, Environmental Geology, 45:132–143.

Mohamed AMO, Hawas Y (2004) Neuro-fuzzy logic model for evaluating water content of sandy soils. Computer Aided Civil and Infrastructure Engineering, 19:343–356.

Mohamed, A.M.O. and Said, R.A. 2004, TDR detection of nonaqueous phase liquids in sandy soils using the eigendecomposition method, Environmental Geology J., 47, 30–37.

Mohamed, A.M.O., and Said, R.A. 2005, Detection of organic pollutants in sandy soils via TDR and eigendecomposition, Journal of Contaminant Hydrology, 76, 235–249.

Press WH, Teukolsky SA, Vetterling WT, Flannery BP (1992) Numerical recipes in C: The art of scientific computing, 2nd Edition, Cambridge University Press, Cambridge, 994p.

Richards W, Seary A (1997) Introduction to eigen analysis of networks, Proceedings INSNA Sunbelt XVII, San Diego, February 13–17, 1–9.

Said RA, Al-Shawawreh NK, Mohamed AMO (2001) A TDR system for subsurface pollutants detection (I): design and modelling, Second Int. Symp. and Workshop on Time Domain Reflectometry for Innovative Geotechnical Applications, Dowding, C.H. (ed.), ©2001 by Infrastructure Technology Institute, Evanston, IL 60201, ISBN 0-9712631-0-8, 361–370.

Sun ZJ, Young GD, McFariane RA, Chambers BM (2000) The effect of soil electrical conductivity on moisture determination using time-domain reflectometry in sandy soil," Can. J. Soil. Sci., 80:13–22.

Vanclooster M, Mallants D, Diels J, Feyen J (1993) Determining local-scale solute transport parameters using time domain reflectometry, J. Hydrol., 148:93–107.

Wraith JM, Comfort SD, Woodbury BL, Inskeep WP (1993) A simplified waveform analysis approach for monitoring solute transport using time-domain reflectometry, Soil. Sci. Soc. Am. J., 57:637–642.

Yong, R.N., Mohamed, A.M.O., and Warkentin, B.P. (1992) Principles of contaminat transport in soils, Elsevier, Amsterdam, 327p.

Arid Land Hydrogeology: In Search of a Solution to a Threatened Resource – Mohamed(ed.)
© 2006 Taylor & Francis Group, London, ISBN 0 415 41127 0

Ways and techniques to protect the desalination plants from oil pollution

K.A. Mohamed & M. Odeh
Water & Power Research Center, Abu Dhabi Water & Electricity Authority, Abu Dhabi, UAE

ABSTRACT: The water resources in the Arab Gulf Region are very limited and do not satisfy the increasing water demands. The desalination of the seawater is the main strategic solution to overcome the water shortage in the region. Large desalination plants were built along the Arabian Gulf Coast. The heavy traffic of the oil tankers in the Arabian Gulf increases the risk of oil spillage which may threaten the desalination plants. Oil protection measures should be installed in the plants and deployed in the contingency events. The early oil spill warning system should be available to provide the plants with the predicted trajectory of the oil slick if it occurs. The warning system will enable the prediction of the arrival time and concentration of oil release when it reaches the plant. The Water & Power Research Center of Abu Dhabi Water and Electricity Authority (ADWEA) has set up oil spill warning system. The paper presents the setup of the developed oil spill warning system in Abu Dhabi. A case study on using the warning system in a real oil spill accident in Abu Dhabi will be presented and discussed.

1 INTRODUCTION

The marine environment is a primary resource in achieving the social, economic, and strategic objectives of the Arabian Gulf Region. The discovery of oil in the region increases the importance of the Gulf due to the essential need of oil around the world. Nowadays, the Gulf is considered as a main source of water for desalination plants which were built along the coast. The desalinated water is used for domestic use in addition to industrial and agricultural purposes.

The discovery of oil fields in the gulf waters makes the marine traffic in the Arabian Gulf very busy. Large oil tankers, sail in the gulf and transport crude oil and other oil products from the Gulf Region to various parts of the world. The heavy traffic of the oil tankers increases the probability of accidental oil spills. The United Arab Emirates is an example of the Arabian Gulf countries where large desalination plants were built along the Arabian Gulf Coast. The Emirates is a federal country consisting of seven emirates located along the western coast of the Arabian Gulf. Abu Dhabi Emirate is one of these emirates and has four large power and desalination plants. More plants will be built in the future to satisfy the water demands.

It is very essential to prevent oil spills, in case they do occur, from reaching any plant intake to prevent significant reduction in the plant efficiency. To achieve this, the Research Center of ADWEA has setup an early oil spill warning system for Abu Dhabi Emirate to assess the risk of the oil spill reaching any of Abu Dhabi power and desalination plants. These plants are Umm Al Nar, Taweelah and Mirfa Power and Desalination Plants. The advantage of this system is to provide an advance warning signal to the responsible authorities so they are able to take the necessary action and to deploy the oil protection measures at the targeted plant intakes. An overview on the developed oil spill warning system and the procedures to be followed to predict the trajectory and concentration of oil slick will be presented. The established procedures by the research center will be clarified when discussing the case study on oil spill accident in Abu Dhabi waters.

2 OIL SPILL WARNING SYSTEM

The oil spill warning system is a numerical transport model. The hydrodynamic forcing imposed in the model is the tidal flow, which is the main hydrodynamic driving force in the Arabian Gulf and Red Sea. The model considers the additional wind drift of the surface slick. A release of the oil spill in the model is distributed over a number of particles; where the mass of each particle represents the amount of a substance attached to it. The number of particles should be specified in the model. Modeling experience dictates that setting the particle number to 100,000 and 400,000 particles for instantaneous and continuous releases, respectively, is a safe assumption. The vertical dispersion of about $0.001\,m^2/s$ should introduced in the model for well mixed flow condition. Practically, the simulation period is between five to seven days. The tidal flow in the oil spill warning system is generated by the relevant hydrodynamic model simulates the tidal movement in the study area. Wind forecast during the simulation period should be introduced in the model to simulate the effect of the wind drift on the slick. The concentration distribution in the slick is computed by the process of spreading (due to gravity, viscosity and surface tension) and the turbulent diffusion. Because time is very critical in the oil contingency it is very important to know as quickly as possible how the slick moves and if it will attack one of the plants. To do the prediction as quick as possible the hydrodynamic models used for the flow simulations should be available and ready for use in case of oil spill occurs.

2.1 *The developed hydrodynamic models in Abu Dhabi*

As mentioned above, the hydrodynamic models are essential for the oil spill warning system and should be available and ready for use in the contingency event. A number of calibrated and verified hydrodynamic models that simulate the Arabian Gulf waters and the waters around the Abu Dhabi power and desalination plants were developed using Delft3D software package of Delft Hydraulics. Figure 1 shows a general layout of the areas covered and simulated by these models.

The hydrodynamic models provide the oil spill warning system with the flow patterns during the simulation period of the oil slick. Selection of the relevant hydrodynamic model for the study depends on the location of the oil release. The Arabian Gulf Model (AGM) simulates the tidal movement

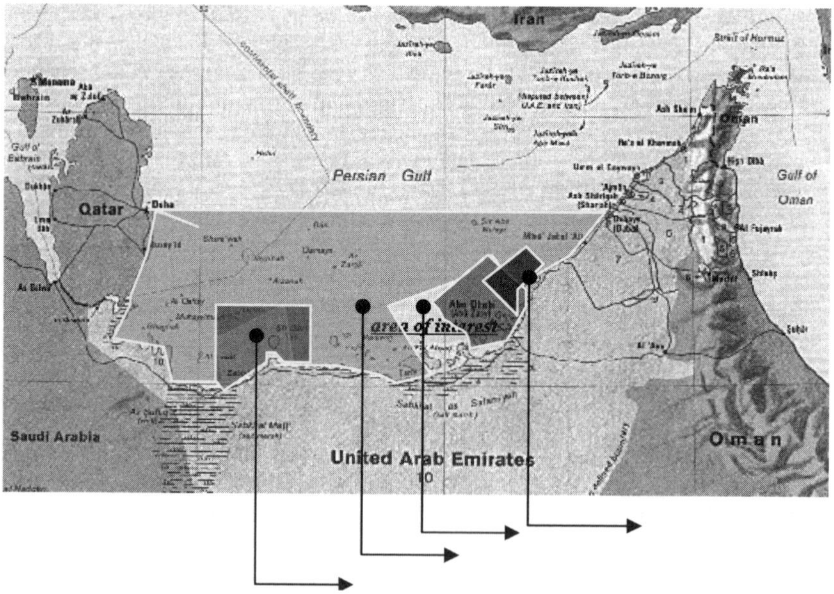

Figure 1. General layout of the developed hydrodynamic models in Abu Dhabi.

in the Arabian Gulf was developed. The Region Model (RGM) simulates the flow patterns in the United Arab Emirates waters were nested in the Arabian Gulf Model. For each one of the Abu Dhabi Power and Desalination plants a hydrodynamic model was developed and simulates the detailed flow pattern in the plant vicinity. Figures 2 and 3 show general layout of the Arabian Gulf (AGM) and the Region (RGM) Models, respectively.

3 THE PROCEDURES OF OIL SPILL MODELING

Once an oil spill accident occurs, the following procedures should be followed to model the spill trajectory and the oil concentration:

Data collection on oil spill: Reliable information on oil spill accident is very important for reliable modeling of the spill trajectory. The data set consists of the location and type of the spill, start time of the release, and the amount of the released oil.
Wind forecast: Wind conditions (speed and direction) influence the advection and transport of oil spill. The wind conditions when the oil is released and wind forecast for about five to seven days

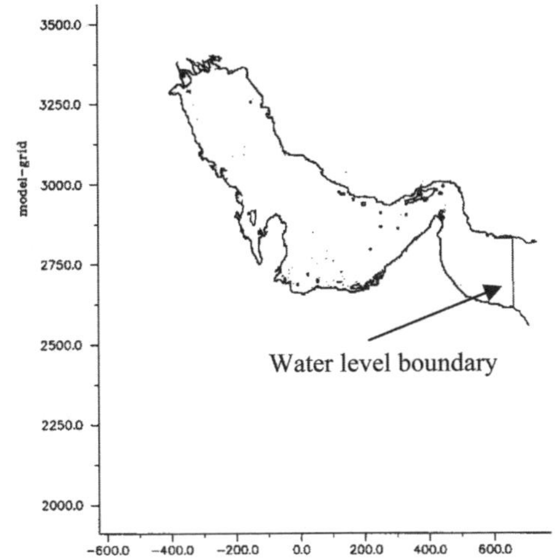

Figure 2. The AGM model.

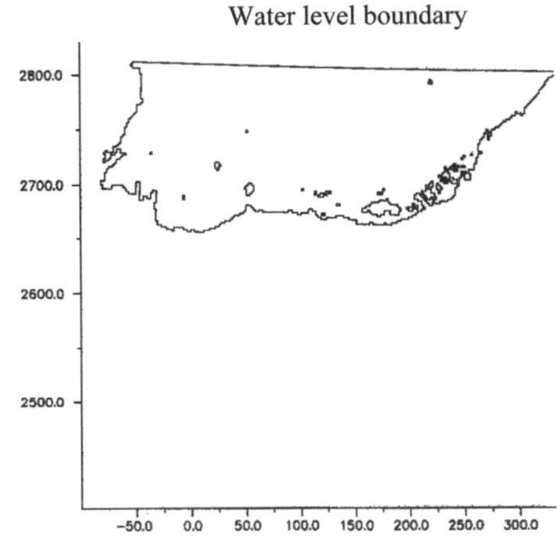

Figure 3. The RGM model.

after the release should be obtained from the meteorological authority and introduced in the oil spill model.

Selection of the relevant hydrodynamic model: The tidal flow, which is important for the transport of oil spills, is simulated with the hydrodynamic model. The location of the oil spill determines the relevant hydrodynamic model to be used to fit the spill in the model grid. If the spill is released in the open sea outside the United Arab Emirates waters the AGM model is used. If it is released inside the United Arab Emirates waters the RGM model is used. From the flow simulation with either the RGM or AGM models a decision will be taken regarding the need for consecutive runs using one of the detailed models developed in the vicinity of the plants. This is done if the simulation shows the spill will approach the shore or the plant.

Hydrodynamic model run: The flow patterns in the study area should be simulated with the selected hydrodynamic model. The simulation period consists of the period as specified in the oil model (5 to 7 days) in addition to two days of model spin up precede the start time of oil release.

Coupling of the flow patterns with the water quality model: The flow patterns generated by the hydrodynamic flow model are then coupled with the oil spill model to enable to use the flow pattern as input in the oil model.

Setup the oil spill model: After the coupling process the other input data and parameters should be introduced in the oil spill model. The start time for the oil release is automatically taken from the hydrodynamic model. The simulation period, time step and the geographical location of oil release should be specified in the model. The spill size presented by the radius of the spill is then determined by the model from the spill volume and the oil viscosity. The amount and rate of oil release should be input in the model. The time series of the wind conditions at the time of release and during the simulation period should be introduced in the model. The evaporation rate is specified based on the oil type. In case of crude or heavy oil the evaporation rate can be set to Zero. In case of light or product oil the evaporation rate can be obtained from oil properties available in the literature or in the oil web sites. The location of the areas of interest (i.e. intakes of the power and desalination plants) should be introduced in the oil spill model as observation points. It will enable obtaining the time series of concentration of the oil slick in order to determine the arrival time of the oil at these locations.

Model results: The output files from the simulation with oil model should be processed and analyzed. Two dimensional plots showing the area of interest and the predicted location of the oil spill at any specified time frame can be obtained. These can then be used to generate an animation of the oil spill trajectory over the simulation period. Time series plots of oil concentration during the simulation period can also be obtained at the specified observation locations. These plots give an indication of when a particular location would be affected and the amount of oil concentration at that point.

Reporting: The results of the oil modeling and conclusions should be summarized in a technical report. Due to the urgent needs a standard format of this report should be made available on the computer to safe time.

The estimated times for the completion of each of the mentioned processes is as follows:

- The estimated time for oil and wind data collection is about one hour after the request for the study is received.
- The selection of the relevant hydrodynamic model takes about half an hour.
- The estimated simulation time of the hydrodynamic model depends on the type of model to be used. Generally, the estimated run time of the hydrodynamic flow model is not more than 45 minutes.
- The estimated time for coupling the flow pattern generated by the hydrodynamic model with the oil model is about 5 minutes.
- The estimated time for the oil spill model setup is about 30 minutes.
- The oil model simulation will take about 15 minutes (on average).
- The estimated time for processing and creating the plots and animation is about one hour.
- The estimated time for issuing the report is about half one hour.

As it can be seen going through the entire procedure, starting from receiving the request for the study until the report is ready will take about 4 to 5 hours. It gives ample time to send a clear warning to the plant which may be attacked by the oil slick.

4 CASE STUDY ON ACTUAL OIL SPILL IN ABU DHABI

On January 24th, 2000 an oil tanker named Ghazeya was sunk offshore Abu Dhabi Emirate causing oil to spill in the Gulf waters. Figure 4 shows the location of this oil spill. Relevant information on oil spill was given by the Abu Dhabi Coast Guard. The Location of the spill was 7 sea miles North East of Abu Dhabi Island at 24° 38.8′ N and 54° 24.3′ E. The time of start of release was on January 24th, 2000 at 10:00 am and the tanker capacity was 900 tons of fuel oil. The wind conditions when the accident occurred as well as the wind forecast for five days after the start of oil release was provided by the Abu Dhabi Meteorological department. The Region Model (RGM) covers the waters of the United Arab Emirates was selected for flow simulation based on the oil spill location. The simulation period of the flow model was 7 days including 2 days for model spin-up. The simulation started on January 22nd, 2000 at 10 am and stopped on January 29th, 2000 at 10 am.

The time step was taken as 3 minutes which gives a good Courant number to guarantee the numerical stability. The flow field from the RGM model was coupled with the oil spill model to provide it with the flow pattern during the simulation period. The geographical location of the oil spill was introduced in the model. The wind drag coefficient was taken as 3% and the wind conditions obtained from the meteorological department were introduced in the model. The evaporation rate was set to 0 because the released oil was heavy oil. The spill size was determined by the model from the release volume and the oil viscosity. The locations of interests were specified in the model. The oil spill model was executed and the output was processed. Figure 5a and 5b show the predicted location of the oil spill after 12 and 24 hours from the start of oil release, respectively.

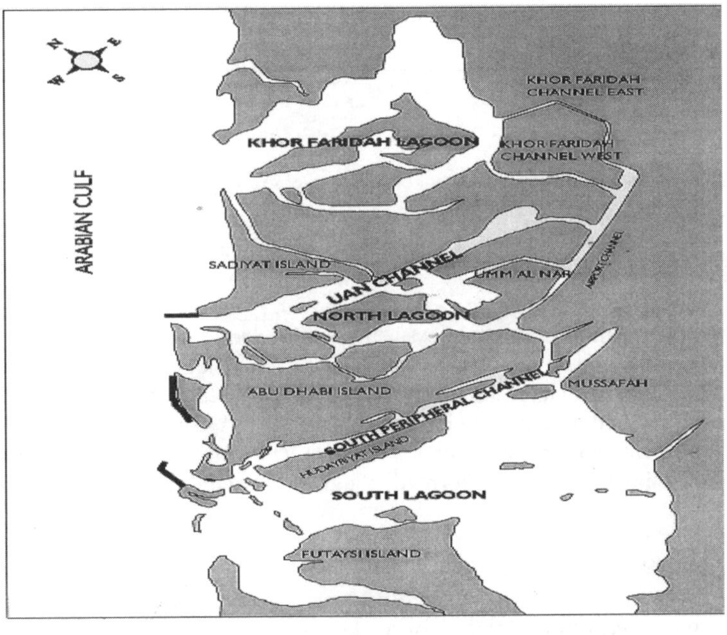

Figure 4. Location of the oil spill.

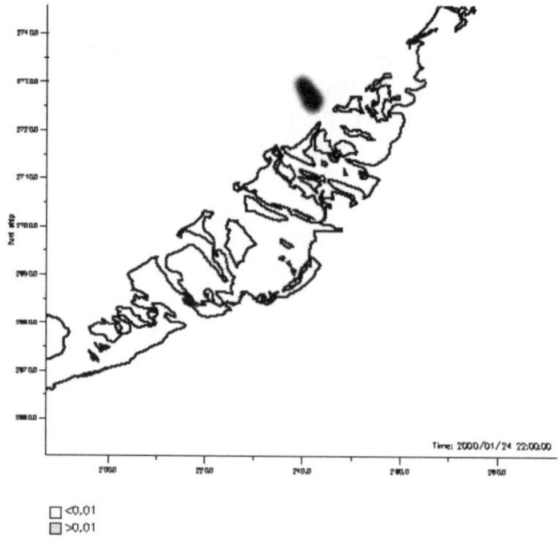

Figure 5a. The prediction of oil spill location 12 hrs after release.

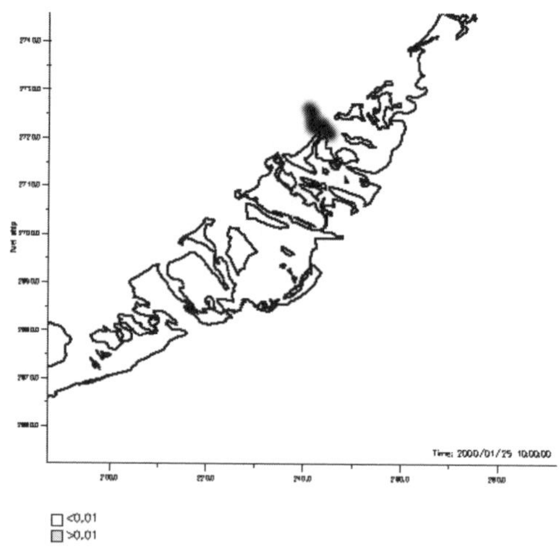

Figure 5b. The prediction of oil spill location 1day after release.

The above figures showed that the slick was closer to Sadiyat Island, after 12 hours from the oil release, and it moved further, hit Sadiyat Island and went through the North lagoon. For the location of the Sadiyat Island and North Lagoon, see Figure 4. Figure 6 shows the time dependent concentration at Sadiyat Island as computed by the oil spill model.

The figure shows that the oil slick reached the island after 20 hours following the oil release. The maximum concentration when it reached the island was about 0.025 kg/m^3. The computational results and the animation covered the simulation period showed that the oil slick did not reach any of the intakes of the Abu Dhabi power and desalination plants. The study was carried out in about 4 hours starting from collecting the relevant information and ending with a report to the officials.

5 CONCLUSIONS

Oil spill modeling is a very powerful tool for predicting the transport and concentration of oil in a contingency event. The oil spill early warning system presented here requires two types of models; a

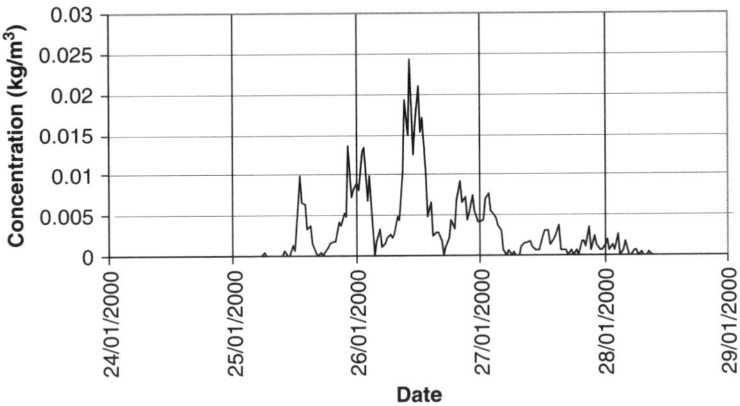

Figure 6. Oil concentration at Sadiyat Island.

hydrodynamic one to describe the flow conditions and the particle dispersion oil spill model that follows the spread of the oil spill accordingly. Wind conditions are important in the prediction of the oil spill trajectory. The effect of wind on the flow pattern is simulated in the hydrodynamic model and the wind drift of the surface slick is simulated in the oil spill model. The accuracy of the wind data is very important for the accuracy of the model prediction. The time for predicting an oil spill trajectory in a contingency event is estimated at about 4 to 5 hours starting from receiving a request for the study until an incident report is ready. This is ample time for the responsible authorities to act accordingly in advance to prevent the oil from reaching strategic locations such as power and water desalination plans' intakes and deploy the oil protection measures, if needed, before the oil slick can reach the plant.

RECOMMENDATIONS

It is recommended to develop a good cooperation between the water and power research centers in the Arab Countries generally and Arabian Gulf Region, specifically. The cooperation between the research centers will allow the exchange of knowledge and experience in the field of desalination plants.

It is strongly recommended to develop a global oil spill warning system in the Arabian Gulf and establish a good network between the research centers in the Arabian Gulf Region. This will enable the prediction of the oil spill trajectory, if it occurs and give warning signal to the important locations along the coast of the Arabian Gulf, which may be attacked by the spill. The oil spill and wind data and the results of the study should be shared between the research centers in the region.

REFERENCES

Gerad A.L. Delvigne and Lamber J.M. Huslen, Simplified Laboratory Measurement of Oil Dispersion Coefficient – Application in Computations of Natural Oil Dispersion, Proc. 17th AMOP Conference, Canada, pp. 173–187, 1994.

Eric Anderson and Roddy Thomas, Oil Spill Modeling – Effective Response & Contingency Planning Tools, Oil Spill Response'98, Singapore, May 98.

Part Module, Technical Manual, Delft Hydraulics, the Netherlands, 2000.

Oil Spill Warning System, Technical Report, ADWEA Research Center, Abu Dhabi water & Electricity Authority, June 2002.

Umm Al Nar Computer Model, Technical Report, ADWEA Research Center, Abu Dhabi water & Electricity Authority, June 2002.

Umm Al Nar Region Model & Detail Model, Technical Report, Delft Hydraulics, the Netherlands, 2000.

Sustainable water resources

Arid Land Hydrogeology: In Search of a Solution to a Threatened Resource – Mohamed(ed.)
© 2006 Taylor & Francis Group, London, ISBN 0 415 41127 0

Using renewable energy sources for water production in arid regions: GCC countries case study

M.A. Dawoud
Monitoring and Exploration, Water Resources Department, Environment Agency-Abu Dhabi, UAE

A.R. Allam, M.A. El Shewey & S.M. Soliman
Research Institute for Groundwater, National Water Research Center, El Kanter El Khairia, Kalubia, Egypt

ABSTRACT: Water resources shortage in GCC countries is the main challenging issue facing the rapid development in various sectors. Recent studies indicated that by 2025 water demand for domestic, agriculture and industrial use in GCC countries is projected to increase by more than 50%. With the rapid development of domestic, industrial, and agricultural water supplies, groundwater resources have been seriously depleted. This overexploitation of groundwater causes a lot of problems and the deterioration of entire ecosystems and thus endangers the livelihood of the entire region. This looming water crisis has boosted the implementation of desalination technologies. All desalination systems are highly energy intensive. Recently, the research community worked intensively on coupling desalination systems with renewable energy technologies in robust and cost effective autonomous small-scale desalination units. Solar photovoltaic, solar thermal and wind energy technologies are particularly advantageous for saving the required power sources for producing desalinated water. These renewable energy resources vary widely in technical and economic characteristics.

This study will investigate the feasibility of implementing desalination technologies in GCC to supplement drinking water supplies especially for remote areas using renewable energy sources. Desalination (or desalinization) is defined as removing salts from brackish groundwater and/or seawater to produce fresh water. Issues critical to implementing desalination technologies are: type of desalination technology, environmental concerns and regulations, energy availability and cost, availability of water source for desalination, and cost to the customer.

1 INTRODUCTION

The Gulf Cooperation Council (GCC) countries comprise six countries: Saudi Arabia, Oman, UAE, Kuwait, Bahrain, and Qatar. This region is arid and its countries already passing the water scarcity line as defined by World Health Organization (WHO) (having renewable water resources less than 1000 m^3/year/capita). The scarcity of renewable water resources and growing gap between demand and available supply of domestic water supply due to rapid growth in population and urbanization is a major challenge (Abdulrazzak, 1995). Currently all the GCC countries use desalination to augment their water supply, either on a national scale or at localized locations.

The desalination market is expected to increase significantly in all these countries as drought conditions worsen, populations grow, and water demand per capita increases due to expansion in industrial activities and development of tourism; all compounded by overall improvement in the standard of living. The latest trend in desalination practice in the GCC countries is to adopt privatization after the successful introduction of independent Power and/or Water Producers (IPP and IWPP) in some countries. Financing schemes via BOO and BOOT systems have been adopted in some of the new desalination projects. Technically, among the newly adopted practices is to build hybrid desalination plants (e.g. MSF-RO in Fujairah in UAE). Normally, in GCC countries dual-purpose plants are built. Recently, however, the demand is more on water than on power, thus leading to the construction of independent desalination plants. The high demand for water has also lead to reduction in the delivery time for newer projects. This time lag has been reduced to less than 24 months and, in some urgent cases, to 12 months.

2 PRESENT WATER RESOURCES SITUATION

2.1 *Groundwater*

Water in the GCC countries is a particularly scarce resource. They are mainly dependent on groundwater, desalination and treated wastewater. Natural water scarcity is further worsened by rapidly growing demands, unsustainable use patterns, increased water pollution and weak management institutions and regulations. Many groundwater aquifers in GCC countries are being mined in an uncontrolled and unplanned manner, either because it has not been possible to regulate the access to these aquifers and/or they are non renewable. Unplanned groundwater mining erodes the economic and social sustainability of the communities that depend on the depleting storage. Whether the groundwater mining is inadvertent or planned, there is a need for guidelines to make the concerned communities better prepared economically and socially to cope with the increasing water stresses as the storage is depleted.

Table 1 shows that the annual groundwater recharge in GCC countries is about 4875 MCM. Groundwater abstractions during 2002 exceeded the annual replenishment of about 14697 MCM which is about 75%. Thus, considerable groundwater mining takes place, mainly for irrigation use. Because of the overexploitation, the actual contribution of groundwater to the total use in the region is more than 75%. At country level, groundwater abstractions are currently the main source of water in the GCC countries. Overall, the contribution of groundwater abstractions to total demand ranges from less than 68% (in Kuwait) to more than 90% (in Bahrain).

By the year 2015, it is estimated that the water deficit in the region will reach about 36% of the renewable plus non-conventional supplies. It is expected that this deficit will have to be met largely from non-renewable groundwater aquifers. Hence, it is important to develop the adequate guidelines and regulations for groundwater mining.

2.2 *Desalinated water production*

In order to meet both the qualitative and quantitative requirements for drinking water standards, domestic water supplies in the GCC countries rely mainly on desalination plants produced water, which are used either directly or blended with groundwater. Rural areas have proposed from a number of desalination plants where water is transported over long distances from coastal areas to

Table 1. Renewable water and groundwater use in the GCC countries (year 2002).

Country	Population (×1000)	Renewable water resources (MCM)			Ground water use (MCM)	GW significance, in terms of	
		Surface water	Ground water	Total		% of renewable GW to total renewable water	% GW use to total demand (year 2000)
Bahrain	677	0.2	100	100.2	258	99.80	91.49
kuwait	2165	0.1	160	160.1	405	99.94	68.64
Oman	2518	918	550	1,468	1644	37.47	89.01
Qatar	599	1.4	85	86.4	185	98.38	53.31
Saudi Arabia	21930	2,230	3,850	6,080	14430	63.32	81.23
UAE	2411	185	130	315	2650	41.27	78.50
Total	30300	3334.7	4875	8209.7	19572	59.43	75.34

interior regions. Desalination of seawater and brackish groundwater will continue to be a viable water supply augmentation option for large number of urban centers in GCC countries. Considering recent cost-cutting innovations in the desalination process, this alternative may prove to be relatively inexpensive supply option in comparison to the development of conventional sources located in remote locations. The availability of desalinated water at relatively low cost may also be an attractive means of meeting industrial water demand because industries have been willing to pay for water at rates higher than domestic and agricultural use rates. In 1990 the GCC countries together produced 1557 MCM annually with a daily rate of about 4.26 MCM or per capita 30 liters a day at the national average (ESCWA, 2001). In 1995 the total annual desalinated water capacity was about 2012 MCM with a total produced water of about 1548 MCM as shown in Table 2 (Bushnak 1995 & Ismail 1995). In order to meet domestic water demands, which is a function of population and urbanization growth, the GCC countries are going ahead with desalination plants construction, despite their relatively enormous costs, which range between 1–1.5 US$/m^3. The total annual desalination capacity of the GCC countries at present (2002) is about 2850 MCM. Many types of desalination processes are used such as MSF, RO, PV, and others. MSF are sued seawater and dominates the desalination market by more than 74%. RO is sued mainly for brackish groundwater treatment.

2.3 *Treated wastewater reuse*

Introduced in the early eighties in most of the GCC countries, treated wastewater represents one of the most important alternatives that can be used to meet some of the present water requirements and to lessen the long term supply vs. demand imbalance faced by these countries. Due to completion of sewage water treatment facilities and urban sewage networks expansion in most of the GCC large cities, relatively large volumes of treated wastewater have become available, and because of environmental considerations, have been treated completely or partially regardless of their utilization. Some of the issues encountered in wastewater treatment and usage in some GCC countries are the low rate of wastewater treatment due to the limited sewage network coverage (around 60% in the

Table 2. Present desalination capacity in GCC countries.

Country	Process	No. of unites	Total capacity (m^3/day)	Capacity (MCM)	Feed water
Bahrain	MSF	8	238630	87.1	Seawater
	RO	16	97534	35.6	Seawater and brackish groundwater
Kuwait	MSF	58	1613973	589.1	Seawater
Oman	MSF	8	161015	58.77	Seawater and brackish groundwater
	RO	14	18992	6.93	Seawater
	ME	1	3000	1.095	Seawater
	VC	4	3115	1.137	Seawater and brackish groundwater
Qatar	MSF	34	499954	182.483	Seawater
	RO	11	10284	3.754	Seawater and brackish groundwater
	ME	1	2542	0.93	Seawater
	VC	11	19590	7.150	Seawater
Saudi Arabia	MSF	85	2560802	934.69	Seawater
	RO	24	346478	126.46	Seawater and brackish groundwater
	MED	1	3780	1.4	Seawater
	VC	1	2000	0.73	Seawater
UAE	MSF	74	2080959	759.55	Seawater
	RO	8	4000	1.46	Seawater and brackish water
	MED	1	141369	51.60	Seawater
Total		360	7808017	2849.94	

Table 3. Treated wastewater production and use in GCC countries (year 2002).

Country	Treated wastewater production (MCM)	Treated wastewater use (MCM)	Used to production ratio (%)
Bahrain	24	24	100
Kuwait	258	250	97
Oman	15	11	73
Qatar	44	44	100
Saudi Arabia	475	225	47
UAE	265	215	81
Total	1081	769	

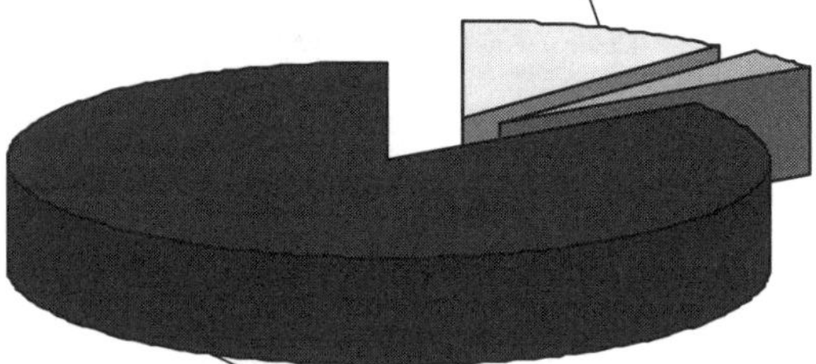

Figure 1. Percentage of water use by source (year 2002).

main metropolitan areas) as a result of the rapid rate of population growth; treatment capacity constraints in the major urban centers that require high investment costs; and the high proportion of wastewater that is treated but not used (Al Zubari 2002). Table 3 displays the current treated volumes of wastewater and the reused volumes in the GCC Countries. At present, all the six countries are operating modern treatment facilities with tertiary and advanced treatment capabilities. The total designed treatment capacity of the major facilities is more than 1,100 Mm^3/y, with a present total treated wastewater volume of more than 915Mm^3/y. However, the recycled volumes of these waters are about 392Mm^3/y, which represents less than 43% of the total treated wastewater. In most of the countries, the remaining unused waters are discharged to the sea. Recycling is used mainly in urban uses (irrigating gardens, roads ornamentals, etc.), fodder crops irrigation, and highways landscaping. Figure 1 shows the percentage of water use by source.

3 WATER DEMAND

Population growth and rapid development in the agriculture and industrial sectors in GCC countries are major issues affecting all sustainable socio-economic development (UNSPD 2002). The estimated population in 2002 was 30.3 million with an average growth rate of 3.73 percent. Population projections for GCC countries over the period 1995–2025 are given in Table 4. The total renewable water in the GCC countries in 2002 amounted to 8209.7 MCM. The high population growth rate in the region exceeds by far the rate of water resource development. Consequently, the annual per capita share of water resources is decreasing, and at an increasing rate. Five countries in the region have a per capita water use of under 500 cubic meters a year, half the benchmark of 1000 cubic meters a year which indicates chronic water scarcity. Saudi Arabia and UAE have done so only by mining their groundwater reserves (Addulrazzak 1999).

Table 4. Past and projected population in GCC countries.

Country	Projected population (31000) 1995	2000	2005	2010	2015	2020	2025	Percentage ration (2025/1995)
Bahrain	557	618	671	717	766	897	1049	188
Kuwait	1691	1966	2192	2390	2576	3076	3673	217
Oman	2027	2717	3302	3986	4752	7002	10316	509
Qatar	548	599	648	693	734	842	967	176
Saudi Arabia	18255	21661	25255	29222	33483	45580	62048	340
UAE	2210	2410	2660	2869	3049	3526	4078	185
Total	25288	29597	34728	39877	45360	60828	81570	323

Table 5. Past and projected water demand in GCC countries (million cubic meters).

	1995			2000			2025		
Country	Domestic	Agri-culture	Industrial	Domestic	Agri-culture	Industrial	Domestic	Agri-culture	Industrial
Bahrain	86	120	17	117	124	26	169	271	169
Kuwait	295	80	8	375	110	105	1100	140	160
Oman	75	1150	5	151	1270	85	630	1500	350
Qatar	76	109	9	190	185	15	230	205	50
Saudi Arabia	1508	14600	192	2350	15000	415	6450	16300	1450
UAE	513	950	27	750	1400	30	1100	2050	50
Total	2553	17009	258	3833	18089	676	9679	20466	2229

Table 6. Water stress index in GCC countries (2002).

Country	Population (×1000)	Renewable water resources (MCM)	Exploited water resources (MCM)	Per capita water resources (m^3/y)	Water stress index (%)
Bahrain	677	100.2	287	164	258
Kuwait	2165	160.1	538	158	156
Oman	2518	1,468	1841	980	74
Qatar	599	86.4	439	466	157
Saudi Arabia	21930	6,080	21155	313	307
UAE	2444	315	3112	316	408
Total	30300	8209.7	25872	358	252

The national economy of most countries depends on oil and oil-related industries, commerce, light industries, and agriculture, in this descending order. Due to the fast increase in population and urbanization, domestic water and industry needs are escalating at rates with which available water resources cannot keep pace. Furthermore, the adopted policy of food self-sufficiency imposes continual constraints on the allocation of water resources, which would otherwise reduce the share for agriculture in favor of increased domestic and industrial demand. Currently the agricultural sector takes 85 per cent of available water resources followed by domestic water use, 14 per cent and 4 per cent for commercial and industrial use as shown in Table 5 (WRI/UNEP/UNDP/WB 2002).

The water stress experienced by the GCC countries in 2002 is expressed in 6 as the percentage of available water resources actually used. The index reaches values of over 100 per cent in five of the six countries and critical values in the remaining one. This indicates that these countries have already exhausted their renewable water resources and are now exploiting non-renewable reserves. The overall value of the water stress index is 252 per cent as shown in Table 6. From this deteriorating water stress index, it is clear that current water resources cannot satisfy future water demand unless positive steps are taken soon to rationalize water demand management, increase and augment supply, and impose realistic controls on use.

4 FUTURE NEEDS FOR DESALINATION

From the analysis of the available data it is clear that the water demand in GCC countries will reach about 32374 MCM by 2025 compared with 24942 at present as shown in Table 7. Desalinated is expected to be the only source to play the main role for augmentation of such deficit due to the limitation of all other renewable sources and the exploitation/deterioration of groundwater aquifers. This furnished a basis for estimating desalination plant capacities required, including those needed to replace existing facilities that have to be scrapped during the same period. Estimation of the capital needed to build these plants is also attempted.

To calculate the future needs for desalination, it was assumed that the desalination will be sued for secure the domestic demand only at 150 liters per capita per day. However, this is not really true and the desalinated water is used in agriculture, parks and amenity plantations but with negligible quantities compared with domestic use. The annual increase in the desalination production by 2025 will be about 1476 MCM which need huge investments as shown in Table 8.

Using the nuclear power for desalination is challenging option to reduce the cost of desalinated water in GCC countries. Although, nuclear reactors are used mainly for production of either heat or power but coupling MSF and RO with nuclear steam supply system can yield economical and technical advantage for producing desalted water. Many studies in Saudi Arabia indicated that the cost of effectiveness of nuclear desalination is a site dependant matter. The type of desalination process and the size and type of nuclear reactor have to be determined based on the specific site data (Al-Mutaz, 2001).

Table 7. Past and projected water demand in GCC countries (million cubic meters).

	1995			2000			2025		
Country	Domestic	Agri-culture	Industrial	Domestic	Agri-culture	Industrial	Domestic	Agri-culture	Industrial
Bahrain	86	120	17	117	124	26	169	271	169
Kuwait	295	80	8	375	110	105	1100	140	160
Oman	75	1150	5	151	1270	85	630	1500	350
Qatar	76	109	9	190	185	15	230	205	50
Saudi Arabia	1508	14600	192	2350	15000	415	6450	16300	1450
UAE	513	950	27	750	1400	30	1100	2050	50
Total	2553	17009	258	3833	18089	676	9679	20466	2229

Table 8. Future desalination needs in GCC countries.

	Quantity (MCM)	
Item	2002 (present)	2025 (future)
Desalination	2849	4325
Exploited groundwater	19572	23450
Treated wastewater	83.6	298
Surface water (from wadies)	2437.2	4300
Total	24941.8	32374

5 USING RENEWABLE ENERGY SOURCES

One of the most the most challenging problem facing the desalination production is the high energy consumption. Efforts are being made to lower the desalination costs. One of the main research fields is using the renewable energy sources for the following reasons:

- Environmental, economic and social sustainability in the energy sector can only be achieved with renewable energies. Present measures are insufficient to achieve that goal.
- A well balanced mix of renewable energy technologies can displace conventional peak-, intermediate and base load electricity and thus prolongs the global availability of fossil fuels for future generations in an environmentally compatible way.
- Renewable energy resources are plentiful and can cope with the growing demand.
- Renewable energies are the least cost option for energy and water security in EU-MENA.
- Renewable energies are the key for socio-economic development and can be used for water production.
- Renewable energies and energy efficiency are the main pillars of environmental compatibility. They need initial public start-up investments but no long-term subsidies like fossil or nuclear energies.

The renewable energy sources can be used for electricity generation and then electricity can be used for seawater desalination by reverse osmosis, while co-generated heat can be applied to multi-effect, vapour compression and multi-stage flash thermal desalination plants. Also combinations are possible. Four categories of renewable angry sources are available in GCC countries namely solar energy, wind energy, hot dry rocks, biomass energy as shown in Figure 2. Table 9 shows the characteristics of contemporary power technologies.

Figure 2. Available renewable energy sources in GCC countries.

Table 9. Some characteristics of contemporary power technologies.

Technology	Unit capacity	Capacity credit	Capacity factor	Resource	Applications	Comment
Wind power	1 kW–5 MW	0–30%	15–50%	kinetic energy of the wind	electricity	fluctuating, supply defined by resource
Photovoltaic	1 W–5 MW	0%	15–25%	direct and diffuse irradiance on a fixed surface tilted with latitude angle	electricity	fluctuating, supply defined by resource
Biomass	1 kW–25 MW	50–90%	40–60%	biogas from the decomposition of organic residues, solid residues and wood	electricity and heat	seasonal fluctuations but good storability, power on demand
Geothermal (Hot dry rock)	25–50 MW	90%	40–90%	heat of hot dry rocks in several 1000 meters depth	electricity and heat	no fluctuations, power on demand
Hydropower	1 kW–100 MW	50–90%	10–90%	kinetic energy and pressure of water streams	electricity	seasonal fluctuation, good storability in dams, used also as pump storage for other sources
Solar chimney	100–200 MW	10 to 70% depending on storage	20 to 70%	Direct and diffuse irradiance on a horizontal plane	electricity	seasonal fluctuations, good storability, base load power
Concentrating solar thermal power	10 kW–20 MW	0 to 90% depending on storage and hybridisation	20 to 90%	Direct irradiance on a surface tracking the sun	electricity and heat	fluctuations are compensated by thermal storage and fuel, power on demand
Gas turbine	0.5–100 MW	90%	10–90%	natural gas, fuel oil	electricity and heat	power on demand
Steam cycle	5–500 MW	90%	40–90%	coal, lignite, fuel oil, natural gas	electricity and heat	power on demand
Nuclear	1000 MW	90%	90%	uranium	electricity and heat	base load power

5.1 *Solar power technologies*

Concentrating solar thermal power technologies (CSP) are based on the concept of concentrating solar radiation to be used for electricity generation within conventional power cycles using steam turbines, gas turbines or Stirling engines. For concentration, most systems use glass mirrors that continuously track the position of the sun. The concentrated sunlight is absorbed on a receiver that is specially designed to reduce heat losses. A fluid flowing through the receiver takes the heat away towards the power cycle, where e.g. high pressure, high temperature steam is generated to drive a turbine. Air, water, oil and molten salt are used as heat transfer fluids. Figure 3 shows the principles of solar thermal co-generation of heat and power. Photovoltaic (PV) cells can be used also for

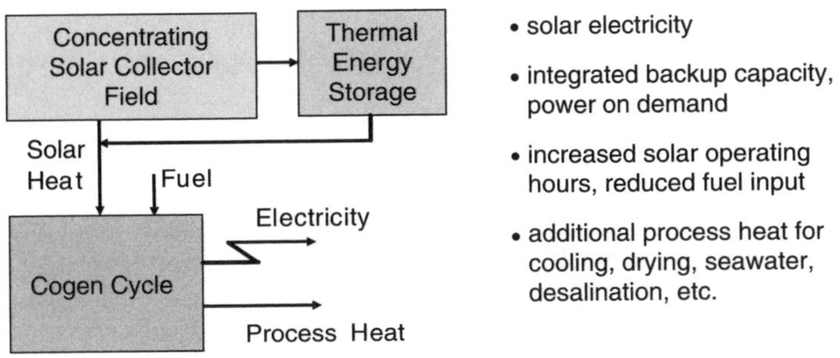

Figure 3. Principles of solar thermal co-generation of heat and power.

electricity generation. PV systems are typically used for distributed or remote power systems with or without connection to the utility grid. Their capacity ranges from a few Watt to several MW. Batteries are usually applied in smaller decentralized supply systems to store the solar energy over the night (Price 2003). There are also scenarios for very large PV systems up to 1.5 GW each to be built in desert areas until 2050.

By 2010, concentrating solar thermal power (CSP) plants are expected to make a significant contribution to the delivery of clean, sustainable energy services in the world's sun belt. A detailed assessment of electricity generation in the Mediterranean region showed a realistic potential by 2020 to 2025 of 23 GW, as compared to an estimated worldwide market of 120 to 140 GW. Furthermore, the assessment indicated a need for activities that would support project development to tackle non-technical barriers and to build awareness of the importance of CSP applications in resolving energy and the environmental problems (Pilkington Solar International 1996).

Two solar desalination technologies have received extensive testing in the GCC countries namely, the solar-thermal MED process and the solar PV-RO process (Knies 2004). The solar-thermal MED demonstration plant which was built in Umm Al Nar near Abu Dhabi city has been in operation since 1984 and is still working at full capacity producing 120 m^3/day of desalinated water using the seawater (ENAA and WED, 1986). The thermal energy required by the MED evaporator was provided by a solar thermal collector field of evacuated tube collectors having an area of 1862 m^2. The plant has shown a high degree of reliability and maintainability. The economics of this technology was shown to be favorable in many remote areas where the cost of fossil fuel is high.

The cost of using solar power technologies is still challenging issue. Several ongoing research activities aim to reach a 20 to 25 per cent energy cost reduction versus conventional oil trough systems. The main improvement targets include: (1) improvements in the collector field as a result of lower-cost designs and more durable receivers and collector structures; (2) development of thermal energy storage systems suitable for solar-only deployment of the technology; (3) continued improvements in the overall operation and management of the systems; (4) system cost reductions and efficiency improvements by substituting water for synthetic oil as the heat-transfer fluid; and (5) development of advanced solar/fossil hybrid designs, especially coupling with combined-cycle power plants. In a dispatchable system, central-station power plants are able to meet the peak load to near-base-load needs of a utility, while a distributed modular plant can serve for both remote and grid-connected application.

The continued technological improvements in CSP systems, along with the cost reductions achieved by system scale-up to larger mass-production rates, have made CSP systems the lowest cost renewable energy in the world. These systems predict cost competitiveness with fossil-fuel plants in the near future, particularly for integrated solar combined cycle systems (ISCCS) that use a mix of solar and fossil fuel resources. Whilst solar power generation costs using CSP systems – solar only – are in the range of 12 to 20 cents per kWh, Solar PACES expects that with continued development and early implementation opportunities, dispatch able system costs could drop to 8 to 10 cents per kWh within five years and 4 to 6 cents per kWh by 2010–2015. Meanwhile, distributed system costs are expected to drop to 12 to 15 cents per kWh within approximately five years and to 5 to 7 cents per kWh by 2010, in the event that reliability problems are solved (Winifred 2000).

5.2 *Wind power*

Wind power can be generated in distributed wind power plants of up to 5 MW capacity each, or in large wind parks interconnecting tens or even hundreds of such plants. There are onshore and offshore wind parks, build into the sea where it is not deeper than 40 m. Wind power is typically fluctuating and cannot be delivered on demand. Wind power is stored for some seconds in the rotating mass of the wind turbines or as chemical or mechanical energy in batteries or large pump storage systems. There are also investigations on storing wind power in form of pressurized air. Fluctuations of the wind velocity are only correlated within a few kilometres of distance. Therefore, the fluctuations of a number of wind mills spread over a large area will usually compensate each other to some extend, leading to power supply transients that are quite manageable by the rest of the power park. However, their share on secured power capacity (capacity credit) is only between 0 and maximum 30% of their installed capacity in very good areas with continuous trade winds (ESCWA 2001). Wind resources are not investigated well in GCC countries. Therefore, appropriate wind resources assessment is recommended.

In UAE only one project has been developed to generate electricity from wind in Abu Dhabi Emirate. The annual power generation 1,875,000 kWh using wind/diesel-system (wind turbine 750 kW and diesel generator set 250 kW) with approximately 2500 full-load h/y and an average annual wind speed of approximately 7.2 m/s in 60 m (Lindemann 2004).

5.3 *Geothermal power (Hot dry rocks)*

Geothermal heat of over 200°C can be delivered from up to 5000 m deep holes to operate organic Rankine cycles or Kalina cycle power machines. Unit sizes are about 1 MW today and limited to about 100 MW maximum in the future. Geothermal energy is often used for the co-generation of heat and power. Geothermal power plants are used all over the world where surface near geothermal hot water or steam sources are available, like in USA, Italy and the Philippines. In the GCC countries those conventional geothermal potentials are small. The Hot Dry Rock technology aims to make geothermal potentials available everywhere, drilling deep holes into the ground to inject cold water and receive hot water from cooling down the hot rocks in the depth. However, this is a very new though promising approach and technical feasibility must still be proven. Geothermal power plants provide power on demand using the ideal storage of the earth's hot interior as reservoir. They can provide peak load, intermediate load or base load electricity. Therefore, the capacity factor of geothermal plants is defined by the load and their operation mode. Assuming a plant availability of 90%, their capacity credit would have that same value (DLR 2005).

5.4 *Biomass power (Waste and Wood)*

There are a number of potential sources to generate energy from biomass: biogas can be produced by the decomposition of organic materials like municipal liquid waste, manure or agricultural residues. Biogas reactors usually require large quantities of water. The calorific value of biogas is about 6 kWh/m^3. Biogas can be used in combustion engines or turbines for electricity generation and for co-generation of heat and power. Landfill gas can be used in a similar way. Solid biomass from agricultural or municipal residues like straw or bagasse and from wood can be used to generate heat and power. From every ton of solid biomass about 1.5 MWh of heat or 0.5 MWh of electricity can be generated in steam cycle power plants. There is also the possibility to raise energy crops. However, this option has been neglected in the GCC countries. Biomass steam-Rankine systems are constrained to relatively small scales because long-distance transport of biomass fuels is costly. As a result, they are generally designed to reduce capital costs at the expense of efficiency.

Table 10. Technical and economic renewable electricity supply side potentials in (TWh/year).

	Solar thermal		Photovoltaic		Wind		Biomass	
Country	Tech.	Econ.	Tech.	Econ.	Tech.	Econ.	Tech.	Econ.
Bahrain	36	33	na	0.3	na	0.1	na	0.2
Kuwait	1525	1525	na	2.5	na	na	na	0.8
Oman	20611	19404	na	4.1	44	8	na	1.1
Qatar	823	792	na	1	na	na	na	0.1
Sadui Arabia	125260	124560	na	13.9	300	20	na	9.9
UAE	2078	1988	na	3	na	na	na	0.7
Total	150333	148302		24.8	344	28.1		12.8

The capital costs per kWe is some \$2000 and the generating cost is in the range of \$0.104 per kWh, including some \$0.012 for biomass collection and transport (Kirvan & Larson 2000).

A variety of technologies are able to convert solid biomass efficiently and cost-competitively into clean and more convenient forms, namely, gases, liquids, or electricity. Most of these are commercially available today. Since desalination plants can be coupled with electricity generation plants in different modes, two categories of biomass technologies for producing electricity can be considered: (1) the first technology burns the biomass from various resources using different combustion systems – conventional or improved fluidized bed. It uses heat to generate steam that drives a conventional steam turbine to produce electricity. This is the most widely used method, and (2) the second approach is to gasify the biomass – gasification – generating a combustible gas that can then be burned, either in a boiler to generate steam or in a gas turbine or piston engine. The capital cost of the gasification route appears to be lower than the cost for a conventional steam generating plant.

Table 10 shows both the technical and economic potentials were defined for each renewable energy resource and for each country. The technical potentials are those which in principle could be accessed for power generation by the present state of the art technology.

6 CONCLUSIONS AND RECOMMENDATIONS

The desalination of brackish water and seawater are becoming more and more important to cover the increasing drinking water demand in semiarid regions especially in GCC countries. The feasibility of these plants supplied by own power stations or the electric grid is indisputable. Energy forms the most important and one of the most costly ingredients of water desalination. Research and development must examine energy issues for desalination that can reduce cost, environmental friendly, improve energy utilization, efficiency and develop new technologies. The following must be considered: (1) hybrid solar and solar/conventional fuel desalination plants; (2) development of energy efficient small desalination systems; (3) assessment of the impact of fuel cell integrated recovery systems and technology on desalination; and (4) innovative alternate energy desalination plants.

The relevant power and water authorities in arid region in general and specially GCC countries must direct efforts towards more accurate evaluation of the possible cost reductions of energy consumed for existing desalination processes by upgrading system efficiency and adopting the off peak desalination concept.

REFERENCES

Abdulrazzak, M.J. 1995. Water supplies versus demand in countries of Arabian Peninsula, ASCE, *Journal of Water Resources Planning and Management* 121(3): 227–234.

Addulrazzak, M. 1999. Role of water legislation in the management of water resources in the ESCWA region", *In Proceedings of 4th Gulf Water Conference*, Bahrain, Vol. 1: 45–63.
Al Zubari, W.K. 2002. Alternative Water Policies for the GCC Countries", *In UNEP Global Environmental Outlook* (GEO) Project.
Al-Mutaz, I.S. 2001. Potential of nuclear desalination in the Arabian Gulf countries. *Desalination* 135: 187–194.
Bushnak, A.A. 1995. Desalination Technology and Economics. *MEC Conference*: Putting a price on Water, Bahrain.
ENAA & WED, 1986. *Cooperation in research and development on desalination using solar energy*. Water and Electricity Department (WED) Progress Report.
ESCWA, 2001. *The role of desalinated water in augmentation of the water supply in selected ESCWA member countries.* Report No. E/ESCWA/ENR/2001/19, United Nation, New York.
German Aerospace Center (DLR), 2005. Concentrating Solar Power for the Mediterranean Region. Institute of Technical Thermodynamics, Section Systems Analysis and Technology Assessment, Final Report.
Ismail, N. 1995, Strategic Projection for Planning and Management of Water Resources in GCC Countries (in Arabic), *Attaawun* 10 (38): 47–62.
Knies, G. 2004. *Solar energy and water for sustainable prosperity in MENA*. Trans-Mediterranean Renewable energy Collaboration TREC, Tripoli, Greece.
Price, H. 2003. *Assessment of Parabolic Trough and Power Tower Solar Technology Cost and Performance Forecasts*. National Renewable Energy Laboratory, Chicago, USA.
UNSPD, 2002. World urbanization prospects. The 2002 Revision (Annex Tables). *United Nations Secretariat, Population Division*, New York, USA.
WRI/UNEP/UNDP/WB, 2002. World Resources 2001–2002", WRI/UNEP/UNDP/World Bank, Oxford University Press, New York, United States, and London, United Kingdom.
Pilkington Solar International, 1996. *Solar thermal power – now*. A proposal for the rapid market introduction of solar thermal technology, Cologne.
Winifred G. 2000. Concentrating solar (thermal) power—the IEA-SolarPACES vision, strategy and activities towards its large scale commercial application. paper presented at the *Expert group Meeting on Disseminating Renewable Energy Technologies in ESCWA Member States*, Beirut (E/ESCWA/ENR/2000/WG.2/4).
ESCWA. 2001. "Overview of wind and biomass systems" Potential and Prospects for Renewable Energy Electricity Generation, Vol. 1, (E/ESCWA/ENR/2001/4).
Kirvan, S. & Larson, E. 2000. Bio-energy Primer: modernized biomass energy for sustainable development, United Nations Development Programme.
Lindemann, J.H. 2004. Wind and solar powered seawater desalination Applied solutions for the Mediterranean, the Middle East and the Gulf Countries. *Desalination* (168): 73–80.

Arid Land Hydrogeology: In Search of a Solution to a Threatened Resource – Mohamed(ed.)
© 2006 Taylor & Francis Group, London, ISBN 0 415 41127 0

Sustainable development: Concept evaluation and protocol development for water resources planners

A.M.O. Mohamed
Department of Civil and Environmental Engineering, UAE University, UAE

ABSTRACT: In this paper, sustainable development was examined from the viewpoints of economical, environmental, social, and land sustainability. A new protocol was developed and discussed. The results of the study have shown that a successful approach to sustainable development must: (1) foster full integration of social and environmental factors along with technical and economic ones from the beginning of the planning process, (2) cross over disciplinary lines and facilitate multi-disciplinary cooperation, (3) produce the analysis and information required to make informed decisions, and (4) leave the subjective societal choices to the decision-makers. In addition, It was apparent that if sustainable development is to be achieved, environmental and social considerations as well as technical and economic criteria must be considered from the very beginning of the planning process by engineers.

1 INTRODUCTION

The need for sustainable development has been recognized for many years. However, it is only within the past few years that it gained prominence as an important concept and philosophy upon which economic development and environmental management should be based. Sustainable development recognizes that environment and development are highly interdependent and should not be dealt with as separate issues. Sustainable development has been defined by various organizations as:

(1) The 1987 World Commission on Environment and Development (WCED) defined sustainable development as:
"*Development that meets the needs of the present without compromising the ability of future generations to meet their own needs.*"

(2) The 1989 Tokyo Conference on Conservation of the Global Environment stated that:
"*Sustainable development calls for a review of not only the conventional framework of the world economy, such as trade, direct investment, international financing and official development aid, but each country's domestic economic, financial and monetary policies.*"

(3) The 1992 Rio Conference on Environment and Development stated, in Principle #3, that:
"*The right to development must be fulfilled so as to equitably meet developmental and environmental needs of present and future generations.*"
and Principle #4 further stated that:
"*In order to achieve sustainable development, environmental protection shall constitute an integral part of the development process and cannot be considered in isolation from it.*"

The basic elements of sustainable development are (Muschett, 1997): (1) population stabilization, (2) new technologies and technology transfer, (3) efficient use of natural resources, (4) waste reduction and pollution prevention, (5) integrated environmental management systems, (6) determining environmental limits, (7) refining market economy, (8) education, (9) perception and attitude changes, and (10) social and cultural changes.

Fundamentally, sustainable development aims for the satisfaction of human needs, the maintenance of ecological integrity, the achievement of equity and social justice and the provision of social self-determination. The real challenge lies in finding ways of putting sustainable development into practice.

2 APPROACHES TO SUSTAINABLE DEVELOPMENT

It is desirable that approaches used by engineers and other disciplines in an effort to achieve sustainable development fit into a multi-disciplinary planning and design environment and have the ability to cross over disciplinary lines and form the common foundation of a unified process. To date, sustainable development has been examined from many different angles by various disciplines, as discussed below.

2.1 *Economics and sustainability*

In view of Principle #4 (UN Rio Conference, Agenda 21, 1992), environmental protection is considered as an integral part of the development process. This is different from the traditional pattern of making economic decisions and then correcting the environmental impacts which may result (Muschett, 1997). This can be illustrated with the use of Figure 1 in which the natural system includes the ambient physical environment, ecosystem and natural resources. The economic system refers to the factors of production of goods and services. Utilization of the natural system by the economic system results in a decrease in the natural resources, and produces additional environmental problems, such as solid wastes, air and water pollution, and greenhouse gases. The importance of these impacts upon the natural system varies geographically, depending on the existing states of both the natural environment and the economy.

2.2 *Environmental sustainability*

Resource scarcity in many areas highlights the need to live within the limitations of the natural environment. We depend on natural capital for resources (inputs) and as a sink for wastes. Environmental sustainability, in the interest of human welfare, seeks to protect the raw materials, and ensure that their capacity as a waste sink is not exceeded Goodland and Daly, 1995). Though non-renewable resource use is by definition unsustainable, one approach is to ensure the depletion rate does not exceed the rate at which renewable substitutes can be created (El Sarafy, 1991). With the maintenance of natural capital approach in mind, Goodland and Daly (1995) draw some practical

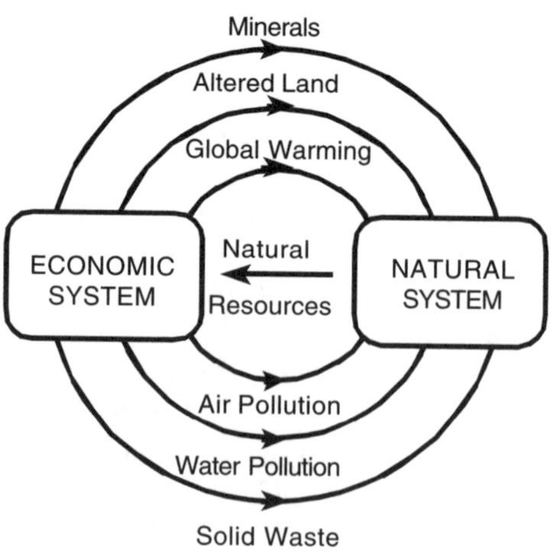

Figure 1. Conceptual interactions between economic and natural systems.

rules of thumb, which they call the input/output rules, to guide economic development. These rules are:

(1) Input rules:
  (a) *Renewable*: harvest rates of renewable resources inputs would be within the regenerative capacity of the natural system that generates them; and
  (b) *Non-renewable*: depletion rates of non-renewable resource inputs should be equal to the rate at which renewable substitutes are developed by human invention and investment. Part of the proceeds from liquidating non-renewable resources should be allocated to research in pursuit of sustainable substitutes (Campbell and Heck, 1997), and
(2) Output rule:
  Waste emissions from a project should be within the assimilative capacity of the local environment to adsorb, without unacceptable degradation of its future waste absorptive capacity or other important services.

2.3 *Social sustainability*

Ways in which social sustainability can be promoted include: (1) encouragement of systematic community participation, (2) emphasizing full cost accounting and *cradle-to-grave* pricing, including social costs, (3) promotion of qualitative improvement of social organization patterns and community well-being over quantitative growth of physical assets, and (4) use of resources in ways that increase equity and social justice while reducing social disruptions.

2.4 *Land sustainability*

Land sustainability can be achieved by integrating the ecological, economical, and social objectives (Holling, 1978). Ecologists stress the preservation of the integrity of the ecological systems that are critical to the overall stability of our global ecosystem, and deal in measurement units of physical, chemical and biological entities. Economists seek to maximize human welfare within the existing capital stock and technologies, and use economic units (i.e., money or perceived value) as a measurement standard. Sociologists emphasize that the key actors in sustainable development are people with a range of needs and desires, and use units which are often intangible, such as well-being and social empowerment.

Sustainable solutions for land development fall at the intersection of the spheres, as shown in Figure 2, that represent the three key ingredients for sustainable development (Campbell and Heck,

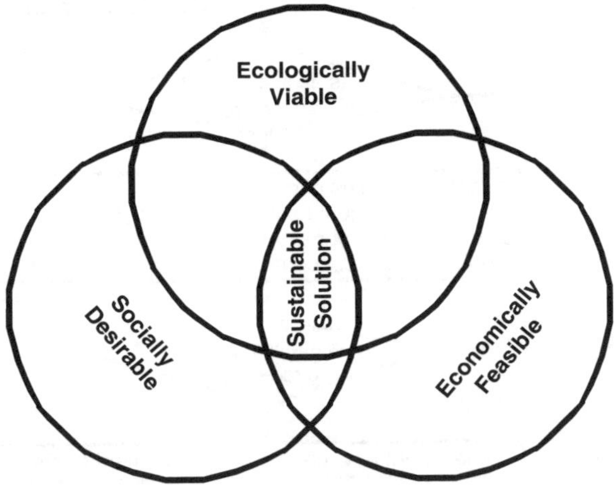

Figure 2. Sustainable solution for the development of sustainable land.

1997). Sustainable development occurs only when management goals and actions are simultaneously *ecologically viable*, *economically feasible*, and *socially desirable*. These imply environmental soundness and political acceptability. Imbalance among the three components, due to failure in one or more of the spheres, will likely result in failure to achieve sustainable development.

3 SUSTAINABLE DEVELOPMENT AND THE AMBIENT ENVIRONMENT

3.1 *Assimilative capacity in environmental management*

The concept of sustainable *yield* was originally developed for harvesting ecological renewable resources, such as forests and fisheries, at the rate at which nature (assisted by human management) was able to replenish. In practice, this is a kind of dynamic equilibrium since natural factors such as climate and ecological productivity vary from year to year. It should also be noted that when the harvest rate exceeds the replenishment rate, a new state of equilibrium with a lower resource base is reached (Muschett, 1997). There are limits to what nature will permit without damaging the ecological system and resource base. Similarly, environmental scientists have come to recognize that the physical, chemical and biological characteristics of the ambient environment determine the ability to accept, dilute, diffuse and transform pollutants. This *assimilative capacity* limits the amount of pollution tolerable without causing damages.

This principle holds whether we are considering a very localized leachate plume from a landfill site or in a river, a regional air pollution problem or a global climate change (Muschett, 1997). In general, as the geographical scale increases, the complexities and interactions of natural processes also increase. Therefore, to achieve sustainable development, we must consider the assimilative capacity of the environmental system, which in turn determines the *carrying capacity* of the supporting population and economic activity and the resulting pollutant emissions (Mohamed and Antia, 1998).

A general framework for environmental quality management is shown in Figure 3 (Muschett, 1997). There is often an iterative process which examines different strategies, the resulting spatial patterns of discharge of pollutants, and the modelled ambient concentrations in terms of the assimilative capacity. Ultimately, one or more environmental strategies are selected and implemented.

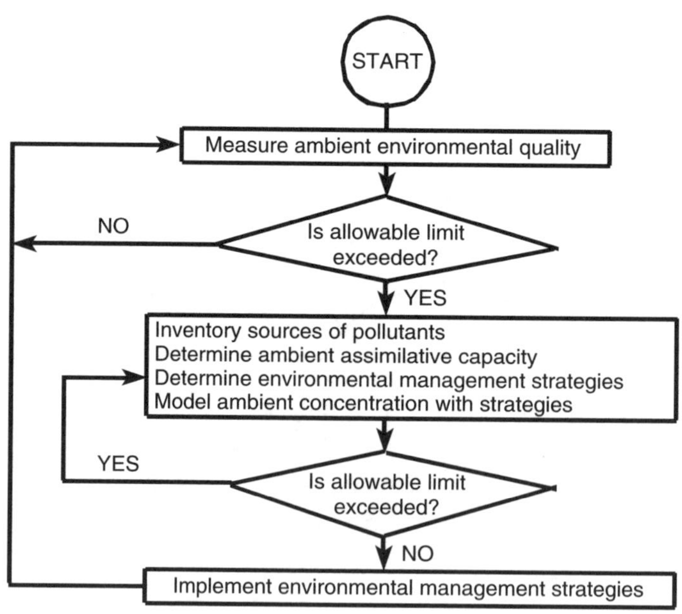

Figure 3. General framework for pollutant management.

3.2 *Water quality management and sustainable development*

Depending upon the amount of pollutant discharged into a water body and the assimilative capacity, the resulting concentrations of the pollutants in the water and in the tissues of aquatic organisms will determine whether the body of water is fit or unfit for human consumption, aquatic life, commercial fishing, recreational purpose or industrial use.

There are two constraints that the environmental system poses in relation to water quality. First, river flows tend to be extremely variable from season to season, and coupled with withdrawals for human use, there is a severe upper limit on assimilative capacity. Second, the increasing bioaccumulation of pollutants in successively higher levels of the food chain also severely limits the allowable concentration of pollutants in the water. The combination of these limiting ambient conditions, together with a high density population and economic activity, can make sustainable development very difficult to achieve.

4 ENGINEERING FOR SUSTAINABLE DEVELOPMENT

If it is accepted that the engineering profession has a responsibility to respect the principles of sustainable development, then every engineer should acquire and maintain an understanding of the goals and issues related to sustainability and conduct his/her work in a manner which supports sustainability. Engineering for sustainable development cannot be achieved by engineers alone, but must evolve within the framework of a larger planning process in which many disciplines and other parties cooperate. The following steps are recommended for the planning and design process leading to projects or activities which may impact on the environment (Mohamed and Antia, 1998; Mohamed, 1998); they are of particular relevance to projects or activities for which an environmental impact assessment may be required by law.

Step 1: Define the objectives

Defining the overall objectives should be the first task brought for discussion. This task involves many stakeholders, such as local communities potentially affected by the project or activity, the proponent, and other businesses. It is also essential to have the involvement of the decision-makers at this point – those who will decide in the end which option best meets the stated objectives. In this process, sustainability must also be declared as a principal objective. Objectives which relate to economic, social and environmental sustainability are (Goodland and Daly, 1995):

(1) *Social objectives*: empowerment, participation, equity, poverty alleviation, social cohesion, population stability, and institutional development;
(2) *Economic objectives*: development for all countries, growth for less developed countries, efficiency, poverty alleviation, and equity;
(3) *Environmental objectives*: ecosystem integrity, conservation of carrying capacity, climatic stability, and conservation of biodiversity.

Step 2: Adopt a cooperative approach

In order to progress effectively through the planning and design process, respective disciplines and stakeholders must be able to cooperate. Adopting a structured cooperative approach can help prevent the process from being paralysed by disciplinary bias or by conflicting views. There are three common approaches used in environmental impact assessment processes for obtaining input from all the relevant disciplines (Mohamed and Antia, 1998): (1) interdisciplinary team, (2) modelling workshop, and (3) study tasks (individual discipline).

Employing such structured techniques in planning and design will foster efficient and productive results from the process. This methodology recommends using one of the first two approaches to benefit fully from the involvement of all participants. When required, the study tasks approach may be used in addition to the primary approach.

Step 3: Develop options to meet objectives

The search for alternative means of fulfilling the objectives should be a multi-disciplinary multi-stakeholder exercise. Individuals that would normally be involved in the analysis of options should also be involved at this earlier stage of defining potential options. Options can arise from efforts of individual disciplines (such as engineers), but can also be created or enriched through the creative exchange between all parties. In this way, engineers involved in preliminary design work will have the benefit of input and feedback from the various key participants.

Step 4: Identify the effects of each option

This step involves analysing the consequences of each alternative design, plan or action. Activities include identifying the benefits as well as the potential environmental and social impacts of each option and developing criteria and indicators for evaluation. This stage may also involve data collection as well as modifications to options and/or the addition of new options. Once different solutions are put forward which meet the objective(s) to varying degrees, discussion should focus on identifying the various effects, both positive and negative. What is often lacking at this point is a common language and a framework which will keep the discussion focused, and which will aid in progressing towards better design and choices.

Knowledge from the different disciplines needs to be integrated and studied on a common ground. A goal should be to define both the sources of environmental impacts as well as the mechanisms involved in their occurrence. This knowledge is essential in order to predict impacts. Using this approach, all environmental impacts are then attributed to some *physical change* resulting from the development. Physical change is defined as a temporary or permanent alteration in the physical (or chemical) environment. The physical environment is comprised of land, water and air, but excludes those elements better described as part of the biological or human environments. Repercussions of physical changes on the biological or human environments are referred to as *impacts*. By linking impacts and physical changes, the causes of these impacts are better understood, and engineers can readily identify problem areas in the design. At this stage, fundamental design choices can still be revisited, and change made to eliminate, reduce or compensate for various impacts.

Step 5: Evaluate the options

This step involves the most difficult task of evaluating the various options in preparation for the selection of the most appropriate course of action. It is important that all relevant disciplines and stakeholders have an input in this important step. What is required at this stage is a consistent framework that will help structure the process toward making complex decisions. Quantitative decision analysis techniques provide this structure and can aid in making trade-offs between competing attributes. The challenge remains in developing a model that will do so without constraining the decision making process by removing or restricting judgements that can be made. Multi-attribute decision analysis and multi-criteria analysis are techniques that can be helpful at this stage.

Step 6: Select the most appropriate option

The responsibility for selecting the most appropriate course of action belongs to the decision-makers. Sustainable development favours democratic, political decision-making that is locally initiated and participatory. The role of engineers and other disciplines should be to provide a range of options which are designed and optimized to meet the objectives as efficiently as possible, to assess the options with respect to the criteria and objectives, and to present the resulting analyses to the decision-makers in such a form as to give a clear picture of the various aspects involved in the ultimate decision.

To make value judgements at any point in this process which goes beyond professional judgement is to transgress the professional's boundary of responsibility. Practitioners need to avoid making choices or value-based decisions which might in some way limit or skew the analysis on which the decision-maker relies when making the ultimate decision.

5 CONCLUDING REMARKS

It is apparent that if sustainable development is to be achieved, and the requirements of the environmental impact assessment process satisfied, environmental and social considerations as well as technical and economic criteria must be considered from the very beginning of the planning process. This will require leadership on the part of the engineering profession, since it has traditionally dominated the planning process, as well as an evolution towards a more multi-disciplinary approach to design that involves the input of biologists, sociologists, anthropologists, etc., at the earliest stages. Engineers should move swiftly to define a new ethic, a new role, new training, and new approaches to the profession.

A successful approach to sustainable development must: (1) foster full integration of social and environmental factors along with technical and economic ones from the beginning of the planning process, (2) cross over disciplinary lines and facilitate multi-disciplinary cooperation, (3) produce the analysis and information required to make informed decisions, and (4) leave the subjective societal choices to the decision-makers.

Sustainable development is a relatively young area of research. The idea of engineering for sustainable development is even more recent. Further research should include an examination of current engineering school curricula to ensure that students are exposed to the central issues, methodologies and thought processes of the humanities and social sciences, as well as course work dealing specifically with issues relating to engineering and sustainable development. A similar professional development course should be designed and made available to practicing engineers.

REFERENCES

Agenda 21, 1992. UN Conference on Environment and Development, Rio de Janeiro,

Campbell, C.L, and Heck, W.W. 1997. An Ecological Perspective on Sustainable Development; *In: Principles of Sustainable Development, St. Lucie Press, Delray Beach, FL, pp. 47–68.*

El Sarafy, S. 1991. The Environment as Capital; *In: Ecological Economics, Costaza, R. (ed.), New York, Columbia University Press, pp. 168–175.*

Goodland, R., and Daly, H. 1997. Environmental Sustainability; *In: Environmental and Social Impact Assessment, Vanclay, F. and Bronstein, D.A. (eds.), John Wiley and Sons, 1997.*

Holling, C.S. 1978. *Adaptive Environmental Assessment and Management; International Institute for Applied System Analysis, John Wiley and Sons, 377 p.*

Mohamed, A.M.O., and Antia, H.E. 1998. *Geoenvironmental Engineering; Elsevier, Amsterdam, 727p.*

Mohamed, A.M.O. 1998. Engineering for Sustainable Development: Development of a Protocol; *2nd International Conference, Role of Engineering Towards Better Environment (Sustainable Development), 12–15 December 1998, Alexandria, Egypt, 1–9 pp. 1998.*

Muschett, F.D., An Integrated Approach to Sustainable Development; *In: Principles of Sustainable Development, St. Lucie Press, Delray Beach, FL, pp. 1–46.*

Arid Land Hydrogeology: In Search of a Solution to a Threatened Resource – Mohamed(ed.)
© 2006 Taylor & Francis Group, London, ISBN 0 415 41127 0

Application of models for the sustainable management of water resources in Wadi Ham, UAE

Salim F. Akram, Ampar V. Shetty & Ahmed S. Al Matri
Ministry of Agriculture and Fisheries, Dubai, UAE

Mohsen M. Sherif
Civil and Environmental Engineering, College of Engineering, UAE University, Al Ain, UAE

ABSTRACT: The sustainability of water resources in UAE is under threat due to the long term drought periods associated with excessive groundwater exploitation. The groundwater exploitation was carried out without considering the actual aquifer recharge. Therefore, quantification of surface water potential and the magnitude of groundwater recharge in the area are necessary for better management of the water resource. In the present study, HEC-HMS was used to quantify the surface water potential and Visual MODFLOW was implemented to assess the groundwater recharge in the study area. The results of surface hydrology modeling and other hydrogeological information were used to develop the input of the groundwater flow model in Wadi Ham area. The available historical groundwater level data were used to calibrate and validate the model for both the steady and transient conditions. The water budget calculations were used to determine the increase in the groundwater recharge due the construction of Wadi Ham dam in 1982. Different scenarios of groundwater levels were simulated based on the probable surface water, recharge and draft conditions. A Strategy for the sustainable use of water resources in the study area was discussed.

1 INTRODUCTION

The proper assessment and management of water resources in the United Arab Emirates would certainly contribute to water conservation, enhancement of the quality of available water and restoration of the depleted groundwater resources in the various aquifers. Evolved from its prime mission and responsibility to develop and sustain the water resources in the country, the Ministry of Agriculture and Fisheries (MAF) has constructed a large number of dams across the main wadis to harvest the rainwater, protect the environmental system and recharge the aquifers. Most of the dams are located in Fujairah and Ras Al Khaimah emirates because of their mountainous nature which allow for the generation of surface water runoff. A better management of water requires the preparation of a hydrologic balance model for the entire watershed as well as for each part of it. The amount of water can be augmented by the use of conservation methods such as artificial groundwater recharge and water harvesting.

The United Arab Emirates lies in the southeastern part of the Arabian peninsula between latitudes 22° 40′ and 26° 00′ North and longitudes 51° 00′ and 56° 00′ East. It is bounded from the north by the Arabian Gulf, on the east by the Sultanate of Oman and the Gulf of Oman and on the south and the west by the Kingdom of Saudi Arabia. The total area of the United Arab Emirates is about 83,600 km^2. Most of the land is desert and is characterized by the predominance of Aeolian Landform System. The geomorphologic features include mountains, gravel plains, sand dunes, coastal zones and drainage basins.

The United Arab Emirates is divided into two distinct zones: the larger low-lying zone and the mountains zone. The first covers over 90% of the country's area, extending from the northwest to the eastern part of the country where it is truncated by-the mountains zone (Al Hamady, 2003). The low-lying zone ranges in altitude from sea level up to 300 meters. Its major part is characterized by the presence of sand dunes which rise gradually from the coastal plain reaching their highest elevation of 250 m above sea level (a.m.s.l). Along the coast of the Arabian Gulf, the low-lying land is punctuated by ancient raised beaches and isolated hills which may reach up to 40 m (a.s.l) in some locations (Baghdady, 1998).

Renewable water resources in the United Arab Emirates are very limited. No surface water in the form of rivers or lakes is available. The rainfall is very scarce, random and infrequent. The average rainfall is in the order of 110 mm/yr. However, this annual rainfall is mostly encountered in few events. Records indicate that the average number of rainfall events per year is in the order of five or less. On the other hand, the rainfall events are generally characterized by their short durations and heavy intensities. Such rain characteristics are quite consistent with the regions classified as drought areas.

Annual rainfall in the Wadi Ham ranges from 3.7 mm to 505.8 mm. Normal rainfall estimated for 24 years is 151 mm with standard deviation 126.8 mm, kurtosis 1.36 and coefficient of asymmetry 1.18. The probability of occurrence of 75% and 50% normal rainfall were estimated as 51 and 64 percent, respectively. It implies that the distribution of rainfall in Wadi Ham is highly scattered and not dependable on annual basis (MAF, UAEU, 2005).

Rainfall is distributed from January to December with maximum occurring during the months of February and March. About 50% of annual rainfall normally occurs during these two months. Monthly rainfall values range from 0 to 184 mm, mean monthly varies from 1.6 mm to 40.6 mm with variation of standard deviation from 2.7 mm to 60.9 mm. The monthly standard deviation exceeds the monthly average precipitation. This reveals that the year-to-year monthly variation in precipitation is quite extreme in the area (MAF, UAEU, 2005).

The study of groundwater resources sustainability in the UAE is of prime concern to the Ministry of Agriculture and Fisheries. Although groundwater may not be suitable, in most cases, for drinking and other potable purposes, it represents the main source for irrigation. About 85% of the total water consumption for irrigation purposes in the UAE is groundwater (IWACO, 1986). Despite the shortage of natural recharge due to the scarcity of rainfall during the last few years and the associated decline in its level, the groundwater is still regarded as a precious resource for meeting agricultural demands in many areas of the country. Therefore, the importance of the current study evolves from the vital role of groundwater resources in the sustainable development in the UAE.

This paper presents an assessment of surface water runoff that would be generated from rainfall in the vicinity of the area of Wadi Ham. The effect of pumping on the groundwater levels in Wadi Ham has also been investigated. The HEC-HMS model was used to assess the surface water runoff in Wadi Ham under different rainfall events. The model parameters were calibrated to simulate observed surface water flow. A good match was obtained between the observed and simulated surface water runoff. MODFLOW was calibrated and validated over a total duration of 16 years. The Model structure that has been adopted for the study is presented in Figure 1. Different scenarios were simulated to assess the sustainability of the groundwater resources in the area of Wadi Ham.

2 SURFACE RUNOFF MODEL

The Hydrologic Modeling System is designed to simulate precipitation-runoff processes of dendritic watershed systems. HEC-HMS provides a variety of options for simulating precipitation-runoff processes. In the basin model, basin elements like reservoir, sources and diversions are neglected from the present basin model. The models of Wadi Ham contain only 5 elements in the basin model. There are 10 hydrologic elements in the Wadi Ham model, made up of 4 subbasins, 3 river reaches, 2 junctions, and 1 sink is considered at the Wadi Ham reservoir, Figure 2.

In the present study, gauge weighting method of precipitation was considered in the meteorological model. The control specifications contain all the information related to the time of events for the model, including the start time and date, stop time and date, and computational time step

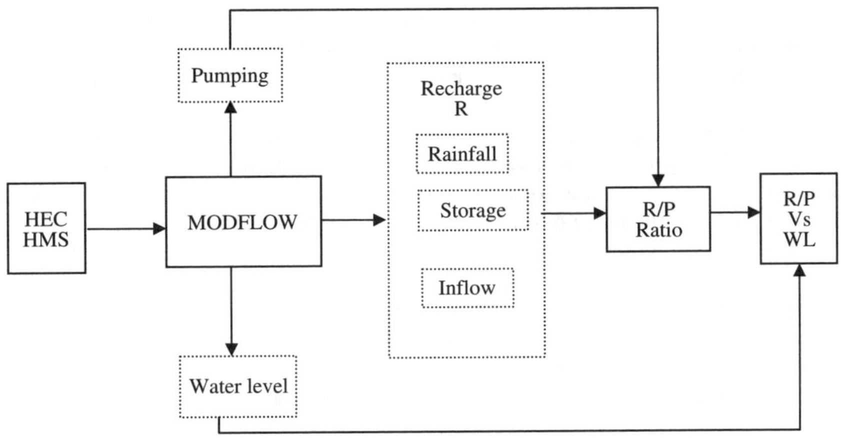

Figure 1. Model structure of the study.

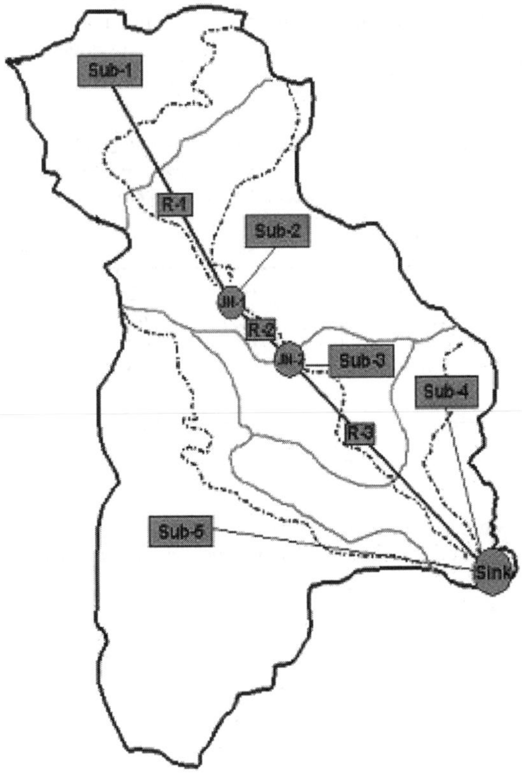

Figure 2. Schematic basin model.

of the simulation. Control specifications were separately selected for all the events of rainfall considered for the simulation.

To perform a hydrologic analysis, raw geometric data including length, slope, area, centroid location, and information such as soil types and landuse/land cover description which are used to characterize the abstractions are manually processed and presented in Table 1. However, other parameters like C_t = basin coefficient, C_p = UH peaking coefficient were selected based on the previous study (Bedient and Huber, 1992) in which C_t typically ranges from 1.8 to 2.2. Also, C_t has been found to vary from 0.4 in mountainous areas to 8.0 along the Gulf of Mexico. It is also reported that C_p ranges from 0.4 to 0.8, where larger values of C_p are associated with smaller values of C_t.

The initial SCS curve numbers (CN) have been selected on the basis of experience and the values quoted in the literature elsewhere (NEH-4; SCS, 1991) based on the hydrologic soil groups and antecedent soil moisture conditions in Wadi Ham. However, CN values were readjusted by trial and error to achieve the best possible results comparable with the observed storage/flow values. Ophiolite formation with steep basin relief was considered in Wadi Ham and the curve numbers ranged between 66 and 69.

The percentage of error between estimated results and the observed measurements ranged between 1 and 32 percent. The error is in the order of 5% for the relatively "important" big storms with water storage of more than 2.0 MCM. The above results indicate that the difference of simulated and observed yields is within the allowable limit except for few cases of small water storage.

2.1 *Estimation of storage in the reservoir*

Different scenarios of incidence of rainfall over the catchment area of Wadi Ham were considered to simulate the storage at the dam site using the calibrated parameters. The simulated storage was used to develop a relationship curve between rainfall and storage. The developed curves are presented in Figure 3. These curves can be used to estimate storage/yield in Wadi Ham under different rainfall

Table 1. Basin model parameters of Wadi Ham (after MAF; UAEU, 2005).

Reach (R1) = 7.8 km Reach (R2) = 6.0 km Reach (R3) = 4.4 km

Sub-basin	Area (Km^2)	CN	C_t	C_p	L (km)	L_c (km)	Snyder lag (hr)
1	30.80	69	0.5	0.7	13.2	4.5	1.28
2	47.02	69	0.5	0.7	12.7	6	1.38
3	19.69	66	0.5	0.7	8.2	4.5	1.11
4	28.34	66	0.5	0.7	10.2	5	1.22
5	69.37	66	0.6	0.7	19.7	8	2.05

CN = SCS Curve number
C_t = Basin coefficient
C_p = UH peaking coefficient
L = Length of the main stream from the outlet to the divide
L_c = Length along the main stream from the outlet to the nearest point to the centriod of the watershed.

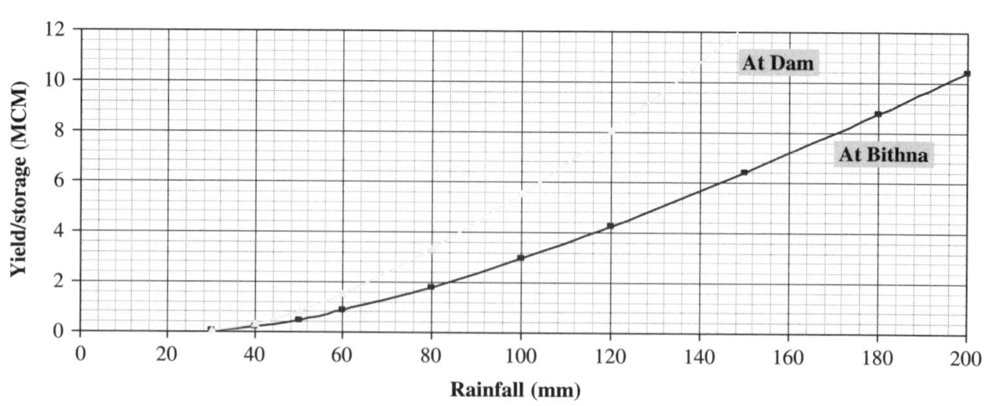

Figure 3. Simulated yield/storage.

intensities. However actual rainfall depth should be derived from the following relationship, which is developed on the weightage of rain gauges for the each sub-basin.

$$\mathrm{R}_{Ham} = (Masafi_{RF})(0.158) + (0.4Masfi_{RF} + 0.6Bihtna_{RF})(0.241) + (Bithna_{RF})(0.101) + (0.7Bithna_{RF} + 0.3Farfar_{RF})(0.145) + (0.35Bithna_{RF} + 0.65Farfar_{RF})(0.355)$$

3 GROUNDWATER FLOW MODEL

Modular three-dimensional finite-difference ground-water model (MODFLOW) was first published in 1984. MODFLOW is a three-dimensional finite-difference ground-water flow model. It has a modular structure that allows it to be easily modified to adapt the code for a particular application. MODFLOW is currently the most used numerical model in the U.S. Geological Survey for groundwater flow problems.

The study domain for Wadi Ham aquifer comprises an area of 117.81 km^2 with total length of 11.9 km east to west (Dam to coast of Oman Gulf) and 9.9 km north to south (Fujairah to Khalba) as shown in Figure 4. The model area and the aquifer boundaries were delineated by digitizing remote sensing image of Wadi Ham. The model domain includes the Gulf of Oman and the ophiolite sequence rock out crops. The ophiolite outcrops are separated as inactive or noflow area. The area of separated outcrop is about 6.56 km^2. At the coast, many cells are located in the sea which is considered to be constant head cells of head 0.0 m (sea level). Ponding area was delineated and marked on the study domain. The total area of ponding at flood level is about 0.40 km^2.

The study area was divided into 119 columns and 99 rows with the size of each cell as 100 m by 100 m. The model is comprised of a total of 11781 equally spaced and square cells. However, the net area of aquifer consisted of only about 64.94 km^2 with 6494 active cells. The area of inactive cells is about 52.87 km^2 and having 5287 cells.

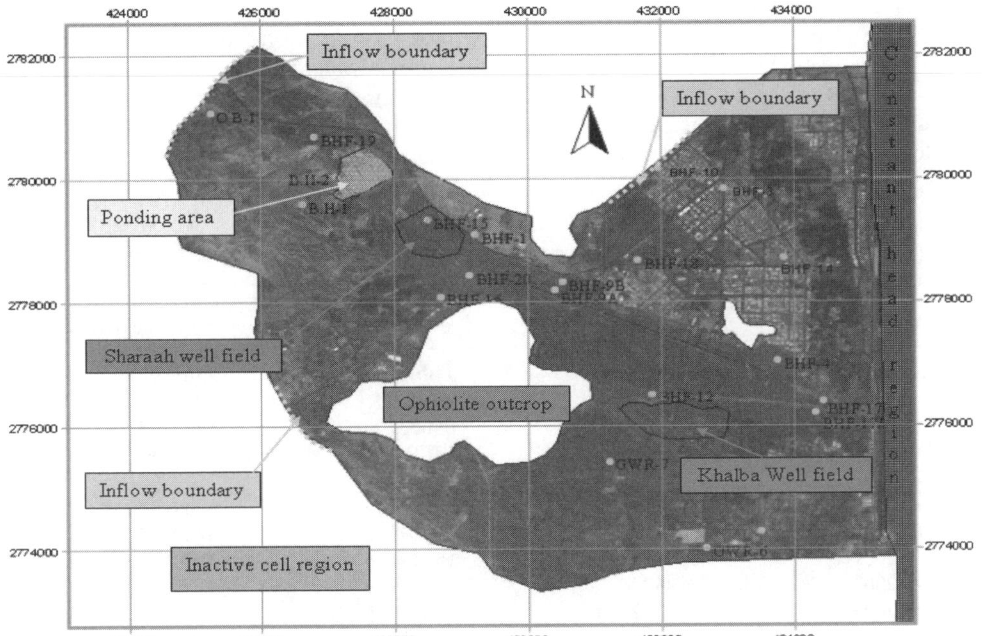

Figure 4. Study domain and boundary conditions.

The calibration period was selected for 5 yrs from January 1989 to December 1993 (1826 days). The length of stress period in this exercise was taken as one real month. However, the period during which recharge of either due to rainfall or dam storage duly considered as an extra stress period. The total number of stress periods considered for the calibration period was 146. The model calibration was achieved by changing three parameters, namely, permeability, specific yield and pumping rates. Abstraction and inflow across the boundaries were also simulated by a number of computer runs till the desirable calculated head in each observation wells was achieved.

The validation period carried out from January 1994 March 2005 for a total of 4108 days. The total number of stress period considered for the validation period was 249. The pumping rate at Sharaah well field during the initial period of validation was about 3150 m^3/day. This rate was gradually decreased down to reach 1700 m^3/day. This represented the closing down of few wells at Sharaah well field.

In Fujairah well field, draft rate scaled down from 2250 m^3/day to 750 m^3/day during the validation period. The pumping from several wells in this field was terminated during the period under consideration. The well field at Kalba experienced a significant the increase in pumping rate from 4000 m^3/day to 20000 m^3/day. The number of wells has increased significantly after 1995. The present pumping rate is considered as 20000 m^3/day.

3.1 *Estimation of recharge factor*

The recharge due to rainfall was also adjusted to ensure that the calculated heads at observation points are reasonably matching the field measurements. The recharge factor is around 20 percent of the rainfall. Although relatively high, the sand and gravel nature of the aquifer system in the study area allows for such recharge. On the other hand, field observations indicated the direct effect of rainfall events on groundwater levels.

The MODFLOW was used to assess the groundwater recharge during the period January 1994 to March 2005. The simulation was conducted under the condition of no dam (no recharge from the ponding area) and the results were compared with the actual conditions including the recharge from the dam. The recharge values after the end of each year are presented in Table 2. At the end of March 2005, the total groundwater recharge due to rainfall and water storage in the ponding was 17495862 m^3. For the other scenario with no water storage in the ponding area, the total recharge was 13969180 m^3. Therefore, the additional groundwater recharge due to the construction of the dam within a period of 11 years and three months was 3526682 m^3. It could be concluded that the dam has increased the groundwater recharge in the area of Wadi Ham by about 20% total recharge and 25 percent of rainfall recharge.

Table 2. Cumulative recharge by rainfall and storage in the reservoir.

	Cumulative recharge (m^3)		
Time (days)	Without storage	With storage	Recharge due to storage
365(Dec1994)	416106	1146295	730189
730(Dec1995)	3972485	5072299	1099814
1096(Dec1996)	7527412	8805219	1277808
1461(Dec1997)	10772370	13659955	2887585
1826(Dec1998)	11904225	15112999	3208774
2191(Dec1999)	12505019	15714176	3209157
2557(Dec2000)	12505019	15714176	3209157
2922(Dec2001)	12505019	15714176	3209157
3287(Dec2002)	12838684	16047949	3209265
3652(Dec2003)	12875759	16085035	3209276
4018(Dec2004)	13470179	16679552	3209373
4108(Mar2005)	13969180	17495862	3526682

The efficiency calculated as the ratio between the total recharge from the storage of dam and the total storage during a specified period is around 47%. Almost half of the water storage is recharging the groundwater system. The remaining 53% of the water storage is mostly held in the unsaturated zone causing a significant increase in the soil moisture content and a part of it may be lost through the evaporation from the ponding area. The estimated storage in the dam and corresponding recharge to the groundwater aquifer is presented in Table 3.

3.2 *Effect of pumping*

The model of Wadi Ham was used to simulate the impacts of abstraction from the well fields in the study domain. The main aim was to understand the behavior of the aquifer system due to the abstraction in the Kalba well field as it is very close to the Gulf of Oman and far away from the Wadi Ham dam. Therefore, the influence of groundwater pumping from this well field on the groundwater levels in the upstream observation wells and the inflow from the constant head boundary was assessed. As an initial simulation, all the pumping wells were switched off and the simulated water table was compared with simulated values with normal pumping over the validation period. The comparison is presented in the Figures 5a–f. These figures show that without pumping, water table levels would have been at a much higher level. The impact of pumping was greater during low recharge periods and low water table levels. The minimum effect of pumping was noticed during the rainy periods. The minimum impact of pumping is observed in wells BHF-4, and BHF-17, which may be attributed to the proximity of Oman Gulf.

Table 3. Estimated storage and recharge.

Rainfall (mm)	Estimated storage (MCM)	Recharge due to storage (MCM)
40	0.2499	0.1175
50	0.8333	0.3917
60	1.4999	0.7049
70	2.4167	1.1358
80	3.2499	1.5275
90	4.4167	2.0758
100	5.5833	2.6242
110	6.6667	3.1333
120	7.9999	3.7599
130	9.2500	4.3475
140	10.8333	5.0917
150	12.0833	5.6791

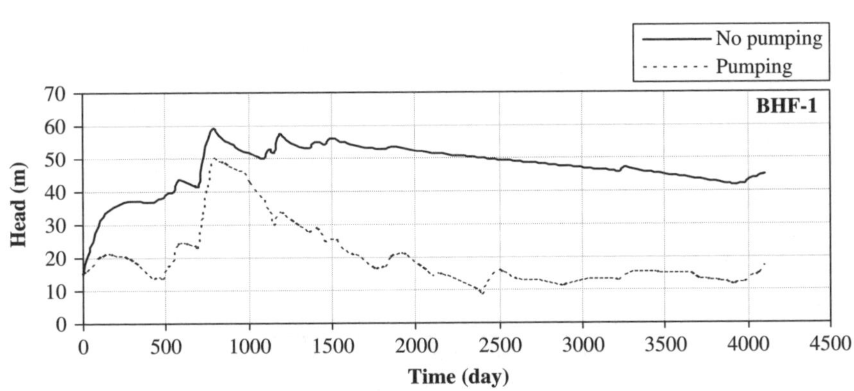

Figure 5a. Comparison of hydrographs of well BHF-1 with and without pumping (amsl).

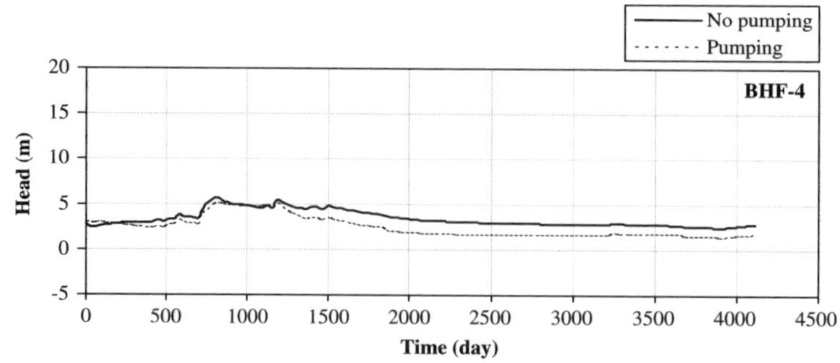

Figure 5b. Comparison of hydrographs of well BHF-4 with and without pumping (amsl).

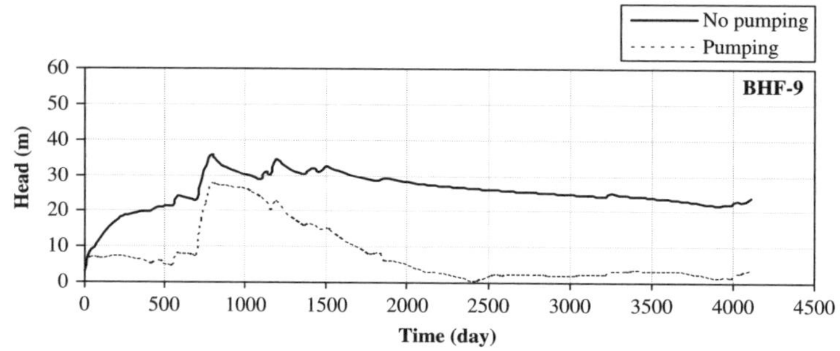

Figure 5c. Comparison of hydrographs of well BHF-9 with and without pumping(amsl).

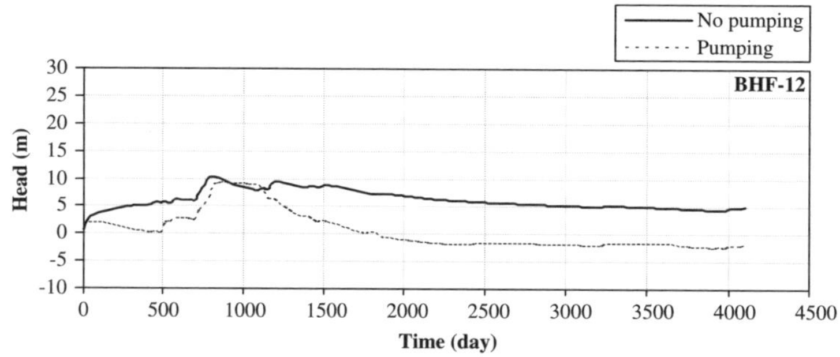

Figure 5d. Comparison of hydrographs of well BHF-12 with and without pumping (amsl).

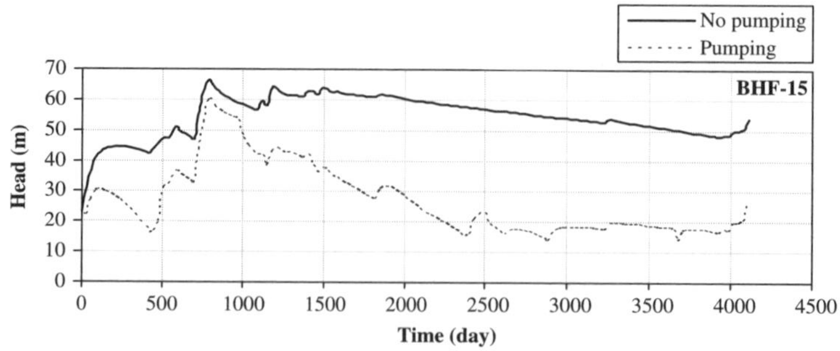

Figure 5e. Comparison of hydrographs of well BHF-15 with and without pumping (amsl).

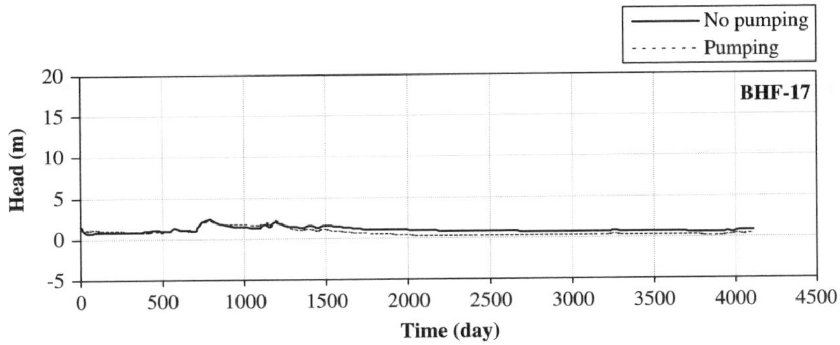

Figure 5f. Comparison of hydrographs of well BHF-17 with and without pumping (amsl).

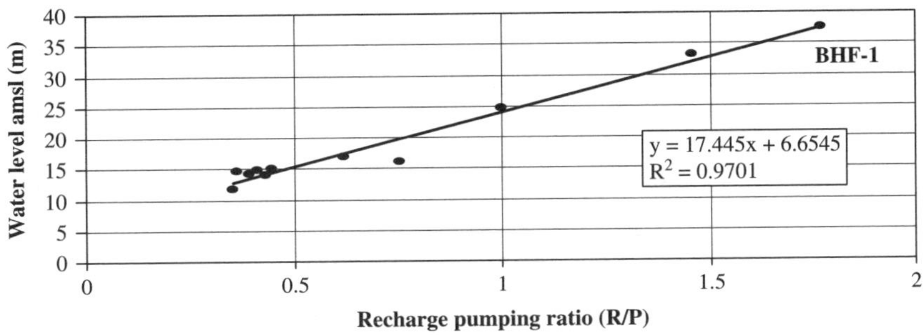

Figure 6a. Recharge-pumping ratio vs. water level at well-BHF-1(amsl).

3.4 *Sustainability of the aquifer*

The stress condition of the aquifer of Wadi Ham is mainly depending on the level of abstraction at Kalba well filed. Normally aquifer under stress when the water level approach the aquifer bottom or attain negative hydraulic head due to different conditions including static, tidal and pumping. Negative hydraulic gradient between the sea shore to Kalba well field should not be permitted. It is advisable to maintain a tangible quantities of groundwater flow toward the shore side to avoid the seawater intrusion and hence deterioration of the groundwater quality.

The groundwater levels in the observation wells represent the impact of recharge and pumping in the area. To understand the water table situation, an effort was made to develop a simple relationship between recharge, pumping and water levels in the aquifer. The MODFLOW was used to estimate the water levels based on the existing recharge and discharge conditions from January 1994 to March 2005. The yearly recharge due to rainfall, storages in the dam and inflow and abstraction of the aquifer were estimated. The ratios of recharge to the abstraction were plotted against the corresponding water table level at the end of year. The developed relationships for all the observation wells are presented in Figures 6a–h. It can be seen from these Figures that all the wells indicated very good correlation between the recharge-pumping ratio and water table.

The recharge-pumping ratio is the main indicator to maintain the water table depth at a required location. The well number BHF-12 is located very close the Kalba well field at which normally water table goes below mean sea level. Therefore, the groundwater level is fixed at well BHF-12 and the corresponding groundwater table at other was estimated wells using the relationship developed as presented in Table 4. The estimated water levels show that most of the wells are not drying up except in wells BHF-15 and BHF-9 by keeping the water levels ranging from −0.5 to 2.0 m in well BHF-12. In other words, most of the water table levels are not reaching the aquifer

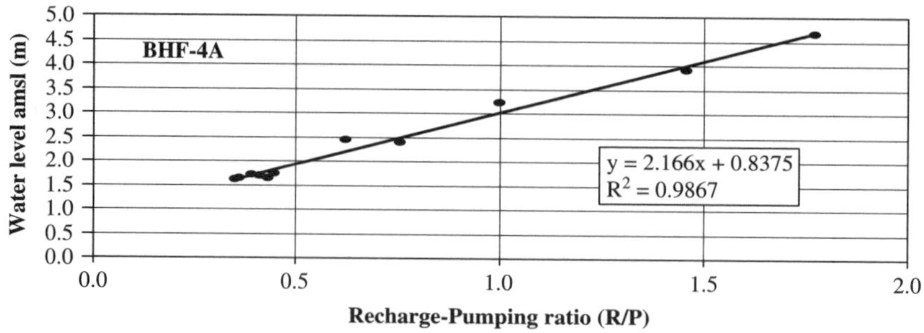

Figure 6b. Recharge-pumping ratio vs. water level at well-BHF-4 (amsl).

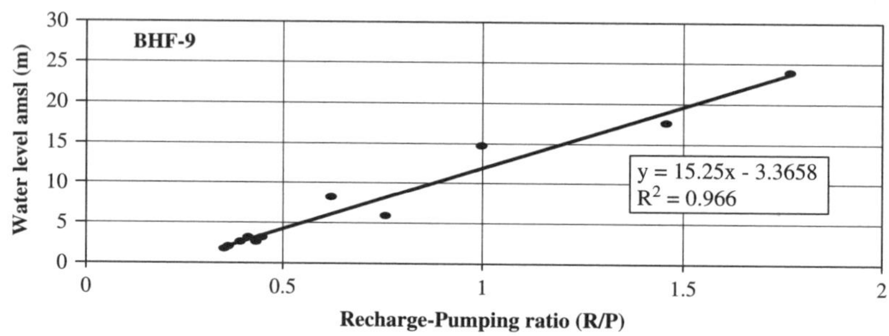

Figure 6c. Recharge-pumping ratio vs. water level at well-BHF-9 (amsl).

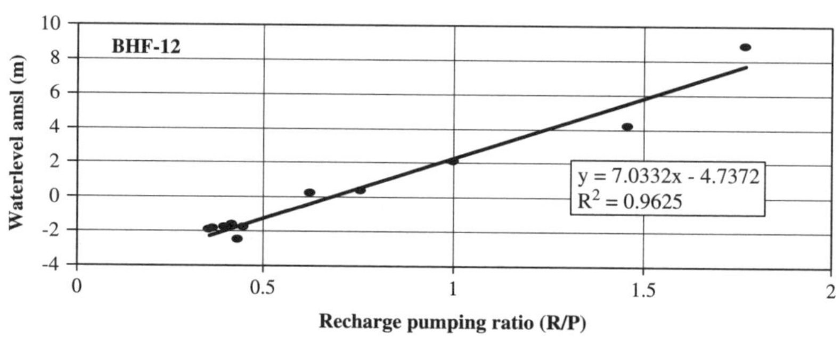

Figure 6d. Recharge-pumping ratio vs. water level at well-BHF-12 (amsl).

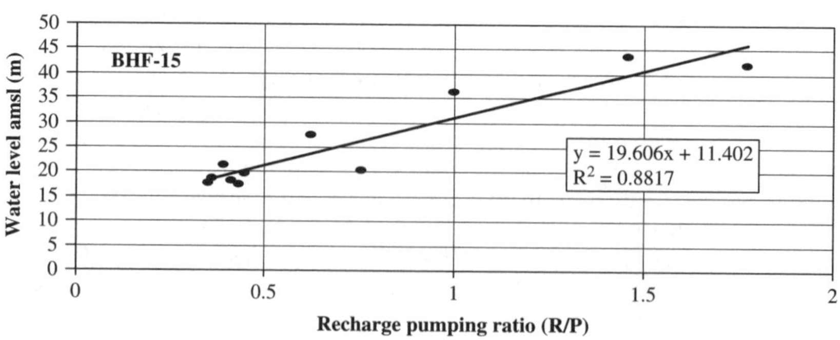

Figure 6e. Recharge-pumping ratio vs. water level at well-BHF-15 (amsl).

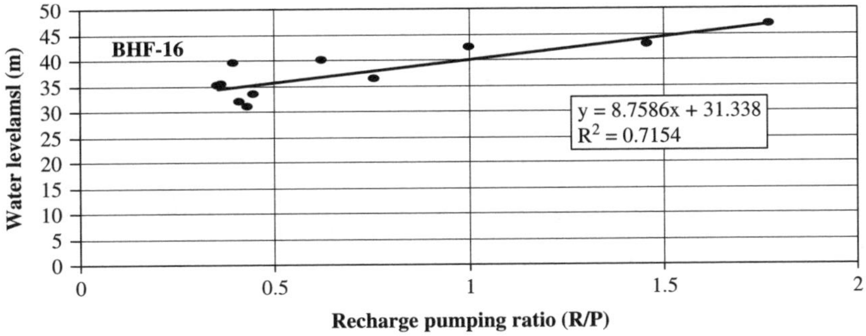

Figure 6f. Recharge-pumping ratio vs. water level at well-BHF-16 (amsl).

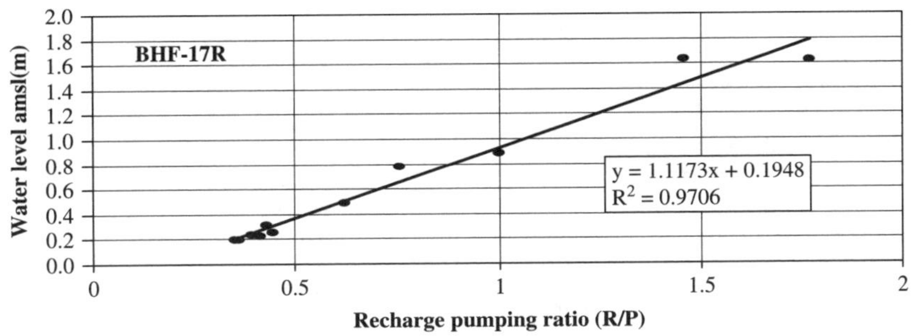

Figure 6g. Recharge-pumping ratio vs. water level at well-BHF-17R (amsl).

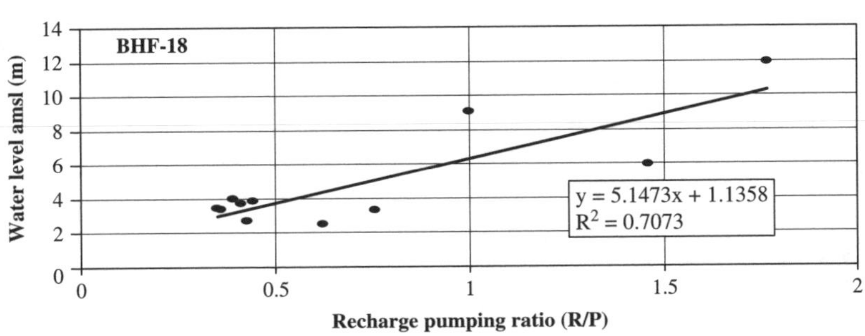

Figure 6h. Recharge-pumping ratio vs. water level at well-BHF-18 (amsl).

bottom. To keep the groundwater levels between −0.5 to 2.0 m, the recharge-pumping ratio was varying between 0.602 and 0.958. However, it is desired to keep the water level at least about 1.0 m at well BHF-12 to keep the positive hydraulic gradient from the well to the sea.

If a minimum groundwater table of 1.0 m and 2.0 m were fixed at BHF-12, then based on the results of the model, the resulted allowable abstraction from the study area is presented in Table 5. In the years 1995, 1996 and 1997 abstraction rates were within the limits of the allowable values. In the other years, the abstraction rates exceeded the allowable limit and caused a stress on the groundwater system. It was clearly observed as the groundwater levels declined to about −2.0 m below the mean sea level in some areas. Figure 7 provides a relationship between the allowable pumping and the total recharge.

Table 4. Estimated water levels in the study area.

		Water level at well BHF-12 (m)					
		−0.5	0.0	0.5	1.0	1.5	2.0
		Recharge-pumping ratio					
		0.602	0.674	0.745	0.816	0.887	0.958
Well no.	Aquifer bottom (m)	Water table level (m)					
BHF-1	0.00	17.16	18.40	19.64	20.88	22.13	23.37
BHF-4	−37.0	2.14	2.30	2.45	2.60	2.76	2.91
BHF-9	−0.40	5.82	6.91	7.99	9.07	10.16	11.24
BHF-15	15.00	23.21	24.61	26.00	27.40	28.79	30.18
BHF-16	14.00	36.61	37.24	37.86	38.48	39.11	39.73
BHF-17	−55.0	0.48	0.56	0.64	0.72	0.80	0.88
BHF-18	−21.0	4.24	4.60	4.97	5.33	5.70	6.07

Table 5. Recharge and allowable pumping.

Year	Recharge (m^3)	Allowable pumping at 1.0 m (m^3)	Allowable pumping at 2.0 m (m^3)	Actual pumping (m^3)
1994	3518796	4312250	3673065	4641000
1995	6577308	8060426	6865666	4505400
1996	9545071	11697391	9963540	5388000
1997	10569250	12952512	11032620	10548252
1998	5632313	6902344	5879241	9029080
1999	3539556	4337691	3694735	8955095
2000	2946322	3610689	3075493	8092450
2001	3021756	3703132	3154234	8548656
2002	3344881	4099119	3491525	7482180
2003	2993586	3668610	3124829	7235838
2004	3135206	3842164	3272658	7252974

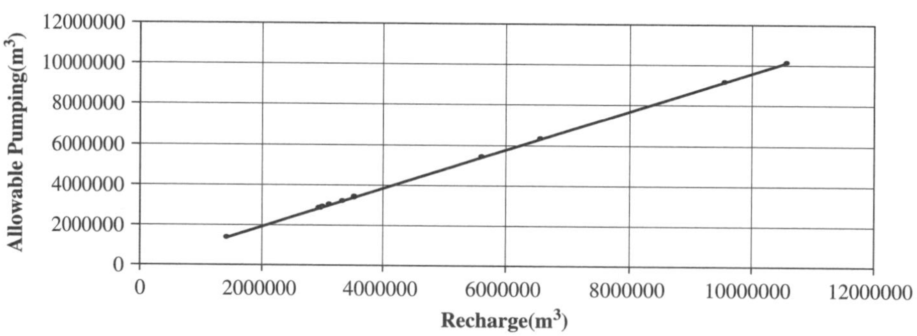

Figure 7. Relationship between allowable pumping and total recharge BHF-12.

4 CONCLUSIONS

The HEC-HMS model was used to estimate the surface water runoff and storage under different rainfall events. The calibrated parameters were employed. The estimated water storage was compared with the observed storage. A relationship was developed to forecast the surface water yield to the dam.

MODFLOW was used to simulate the groundwater conditions in the aquifer of Wadi Ham. The model parameters were calibrated for a duration of 5 years. The validation period extended for another 11 years. A good match was obtained between observed and simulated groundwater levels in the observation levels. The water balance components were obtained from the validation period. Groundwater abstraction has a major impact on the status of the groundwater condition. The yearly recharge due to rainfall, storage and inflow across the study area and pumping were considered to establish a relationship between recharge-pumping ratio and water levels in the observation wells. The developed relationship can be used to define the required water level based on the recharge status in a year. To maintain a 1.0 m water level above the sea at well BHF-12, it is required to keep recharge-pumping ration at 0.816. Therefore, by the known recharge component, pumping can be estimated. For the sustainable management of the aquifer, pumping can be restricted to the allowable pumping estimated by the model. Quality aspect may also be taken into consideration while fixing the minimum water table at a particular location.

REFERENCES

Al Hammadi, M.K., 2003. Assessment of Groundwater Resources using Remote Sensing and GIS, M.Sc. Thesis, Water Resources Program, UAE University, UAE.

Baghdady, A.R., 1998. Petrography, Mineralogy and Environmental Impacts of the Sand Dune Fields of the Greater Al Ain area, United Arab Emirates, Ph.D. Thesis, Faculty of Science, Ain Shams University, Egypt.

Bedient, Philip B. and Huber, Wayne C. 1992 Hydrology and Floodplain Analysis, Second Edition. Addison Wesley.

IWACO, 1986. Groundwater study. Drilling of deep water wells at various locations in the UAE.

MAF., 2001. Climatological data, Vol. 4, 1992–93 to 1999–2000, Water and Soil Department, MAF, UAE.

MAE and UAEU. 2005. Assessment of the effectiveness of Al Bih, Al Tawiyean and Ham Dams in groundwater recharge using numerical models. Vol.4.

Soil Conservation Service, 1971. National engineering hand book, section-4: Hydrology, USDA, Springfield, VA.

Soil enhancement

Arid Land Hydrogeology: In Search of a Solution to a Threatened Resource – Mohamed(ed.)
© 2006 Taylor & Francis Group, London, ISBN 0 415 41127 0

Tubular solar desalination and improvement of soil moisture retention by date palm

T. Fukuhara & Kh. Md. S. Islam
University of Fukui, Fukui-city, Japan

ABSTRACT: This paper consists of two different studies. At first, a Tubular Solar Still (TSS) is proposed to easily remove accumulated salt, which have been the main drawbacks in the basin-type still. To provide better understanding of the mechanism in the distilled water production of the TSS, a field experiment was carried out in the United Arab Emirates in 2002. This paper describes the monthly and diurnal variations of production and the effects of evaporation and condensation on the productivity of the still. Secondly, new use of date-palm as mulch is proposed to improve the soil moisture retention capacity from a viewpoint of the application of unused materials to agriculture in the Middle East. Therefore, heat and moisture transfer experiments were conducted using soil columns and date-palm chips. This paper also describes the performance of the mulch, i.e. evaporation control and suppression of soil-temperature rise associated with incoming short-wave.

1 PRODUCTION OF TUBULAR SOLAR DISTILLATION IN UAE

1.1 *Research purpose*

Due to environmental, ecological and economical activities, particularly in an arid region (such as Middle East) and remote areas, fresh water demand mainly depends on desalination performance produced from brackish or sea water. However, such region is rich in solar energy with high intensity so that solar desalination may be an ideal solution. Solar distillation is the simplest desalination technique, compared to others such as multiple-effect distillation, multi-stage flash, reverse osmosis, electro-dialysis and biological treatment. A basin-type solar still is the most popular method of solar distillation but has undergone very little advances due to low distillate productivity and the difficulty of rapid and easy removal of salt accumulation in the basin. To overcome such difficulties, a new type of solar distillation, Tubular Solar Still designed by Fukuhara et al. (2003).

To achieve better understanding of the distilled water production mechanism of the TSS, this system has been tested in the United Arab Emirates (UAE) since 2001. Temperature and relative humidity in the TSS along with meteorological data were measured in this test.

This paper describes the monthly and diurnal variations of production and examines the effects of evaporation of saline water surface in a trough in the TSS and condensation on the tubular cover inner surface on the productivity of the TSS.

1.2 *Field experiment of Tubular Solar Still*

Field experiments on the TSS have been carried out since September 2001, at the Hamuraniyah farm in Ras Al Khaimah Emirate in the UAE. Figure 1 shows a flow diagram of the experimental setup, which consisted of three water supply tanks, six vessels for collection of distilled water, meteorological observation devices and thirty TSS's assembled in the UAE.

The schematic diagram of the TSS is shown in Figure 2. The tubular cover is made of a curled transparent vinyl chloride sheet of 0.5 mm in thickness and a transparent polyvinyl chloride bottle. The tubular cover is 1.26 m in length and has an outside diameter of 0.13 m. The black trough for

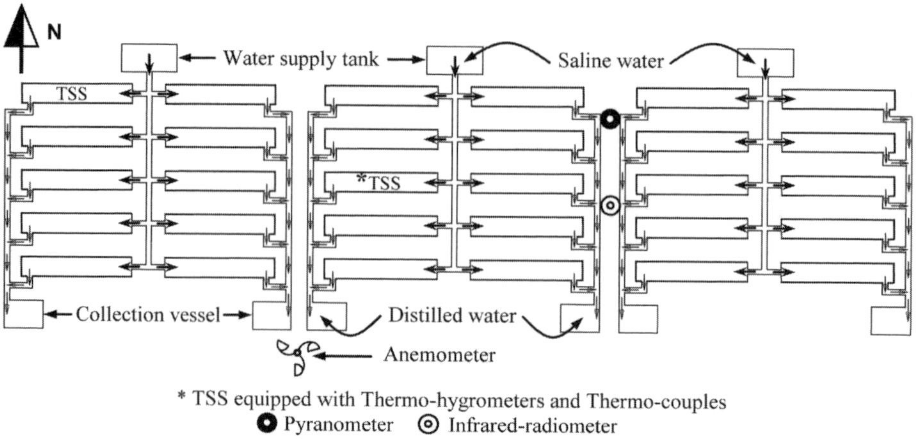

Figure 1. Flow diagram of experimental setup at the Hamuraniyah farm in RAS Al Khaimah, UAE.

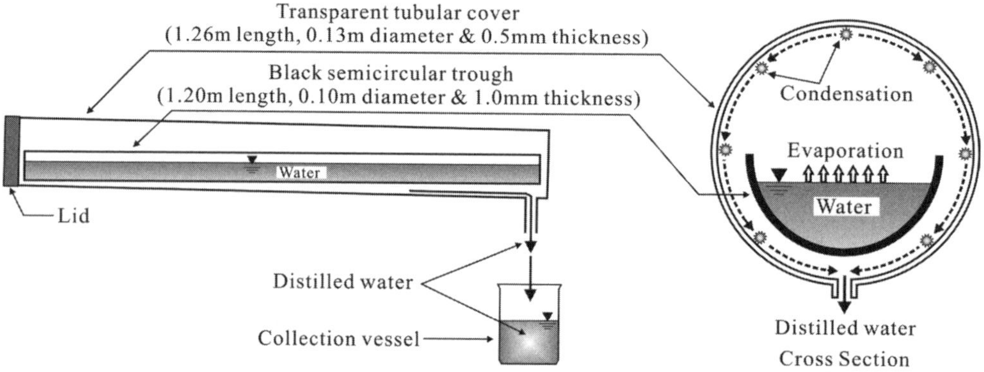

Figure 2. Schematic diagram of Tubular Solar Still.

storing saline water in the TSS is 1.0 mm in thickness, 0.1 m in outside diameter and 1.2 m in length. The attached lid at the end of the tubular cover is easily screwed off and the trough can be promptly taken out and easily inserted back after flushing accumulated salt. During tests, the saline water was, in fact, brackish groundwater from the farm, which contained a salt concentration of nearly 2000 ppm.

Daily measurements of production were made at 7:00 using an electric balance with a minimum reading of 0.1 g and then saline water was supplied to the trough to maintain the initial weight of 1.5 kg. Occasionally, the production was also measured on the hour from 9:00 to 20:00 to obtain the hourly production.

In this experiment, a pyranometer was used to measure the solar radiation, R_s. An anemometer, a thermocouple and a thermo-hygrometer were used to measure the atmospheric air velocity, V_a, temperature, T_a and relative humidity, RH_a, respectively. Furthermore, temperatures of saline-water, T_w, humid air in the TSS, T_{ha}, tubular cover inner surface, T_{ci} and relative humidity of the humid air, RH_{ha}, were measured in one of the TSS's using thermocouples and a thermo-hygrometer, respectively. The data were automatically recorded in a data logger at 30-minute intervals.

1.3 *Experimental results*

1.3.1 *Monthly and daily production*

Figure 3 shows the monthly ensemble averaged daily variations of the production per unit surface area of the saline water in the trough, $\overline{m}_{pd}$, solar radiation, $\overline{R}_{sd}$, saline water and outside air temperatures

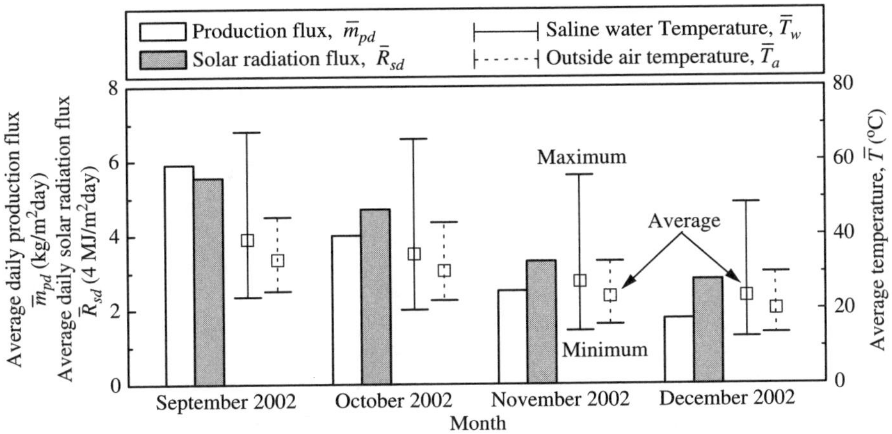

Figure 3. Variations of ensemble averaged daily production flux, solar radiation flux and temperatures per month from September to December in 2002.

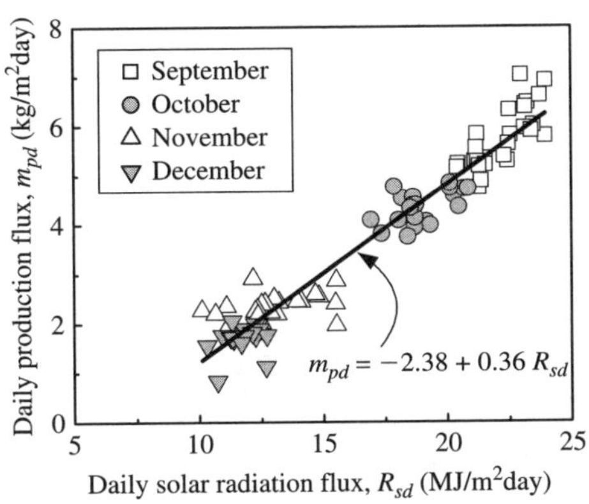

Figure 4. Relation between daily production flux, m_{pd} and daily solar radiation flux, R_{sd} from September to December in 2002.

(maximum, minimum and average), $\overline{T}_w$ and $\overline{T}_a$, respectively, for the period September to December in 2002. The variants $\overline{m}_{pd}$, $\overline{R}_{sd}$, $\overline{T}_w$ and $\overline{T}_a$ were maximized in September and then monotonously decreased till December. The values of $\overline{m}_{pd}$, $\overline{R}_{sd}$, $\overline{T}_w$ and $\overline{T}_a$ were 5.92 kg/m² day, 22.3 MJ/m²day, 39.2°C and 33.1°C in September, respectively, but maximum T_w reached 66.9°C. These values in December were 1.70 kg/m²day, 11.2 MJ/m²day, 23.7°C and 20.4°C, respectively.

Figure 4 shows the relation between daily production flux, m_{pd} and daily solar radiation flux, R_{sd} from September to December in 2002. It can be seen that the value of m_{pd} almost varies linearly with R_{sd}. The correlation between the two variants may be expressed by the regression in Equation (1.1).

$$m_{pd} = -2.38 + 0.36R_{sd} \tag{1.1}$$

It is clear that radiation is one of the main factors that affect the productivity of the TSS.

1.3.2 *Diurnal variation of production*

Figures 5(a) and (b) show the diurnal variations of hourly production per unit surface area of saline water, m_{ph}, solar radiation, R_s, relative humidity of humid air, RH_{ha}, saline water temperature, T_w,

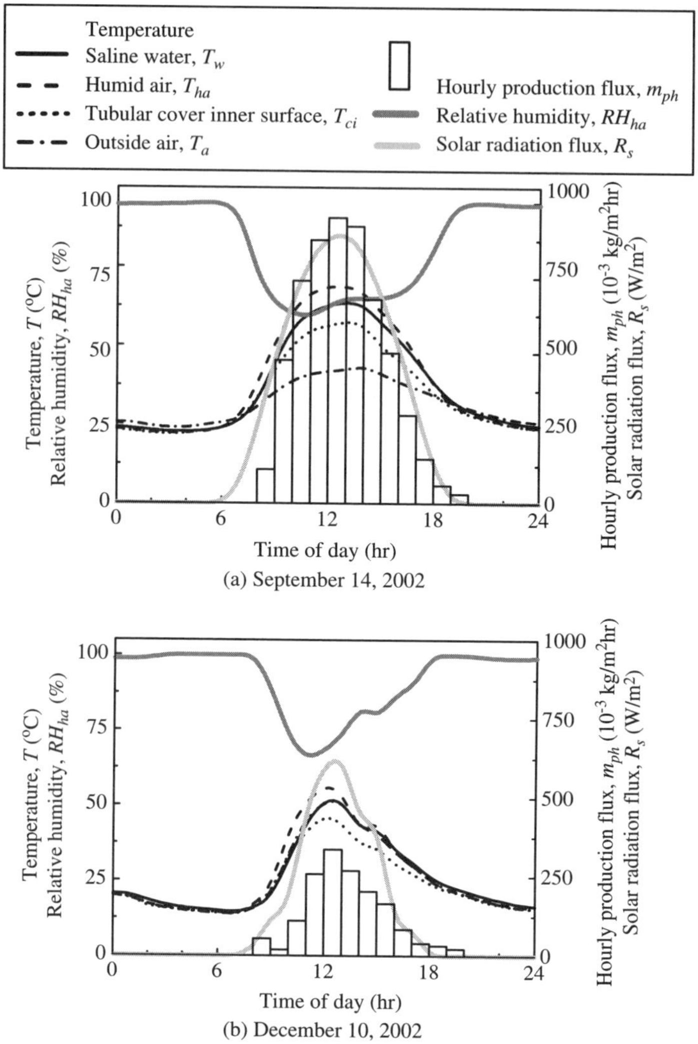

Figure 5. Diurnal variations of production flux, solar radiation flux, relative humidity and temperatures.

humid air temperature, T_{ha}, tubular cover inner surface temperature, T_{ci}, and outside air temperature, T_a, on September 14, 2002 and December 10, 2002, respectively. The values of m_{ph}, R_s, T_w, T_{ci} and T_a rose rapidly after sunrise and peaked between 12:00 and 13:00 before declining gradually. A rapid increase in m_{ph} after sunrise was more prominent in September than December. RH_{ha} was remarkably below 100% during daytime (minimum about 60%) and almost 100% throughout the nighttime. Measured values of the hourly production between 21:00 and 8:00 on the following mornings were negligible. The values of T_w, T_{ci}, T_a, m_{ph} and R_s were higher on September 14 than those on December 10. Similar diurnal distributions were, however, appeared over these four months, not shown here.

1.3.3 *Production mechanism*

Figures 6 (a) and (b) show the diurnal variations of the vapor density on the saline water surface, ρ_{vw}, the humid air vapor density, ρ_{vha}, and the saturated vapor density on the tubular cover inner surface, ρ_{vci}, on September 14, 2002 and December 10, 2002, respectively. Vapor densities ρ_{vw},

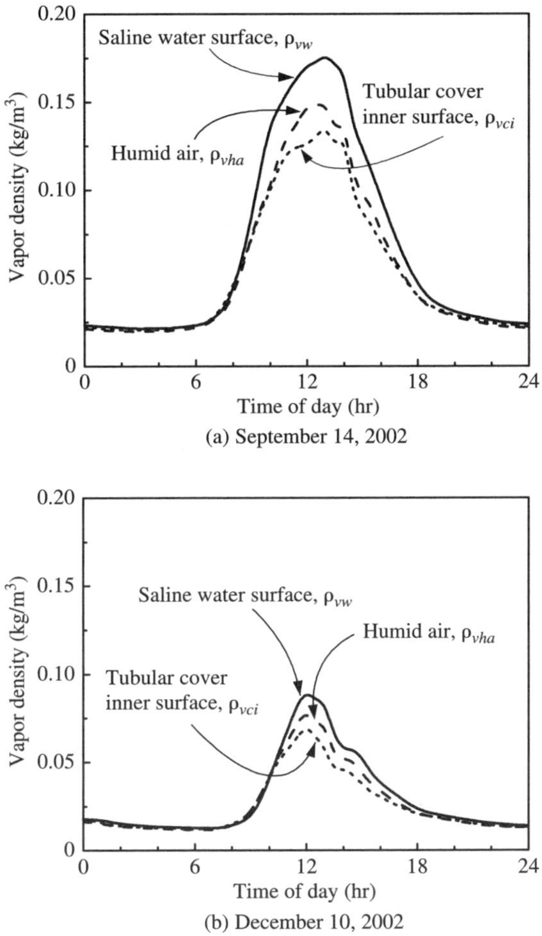

Figure 6. Diurnal variations of vapor densities in the TSS.

ρ_{vha} and ρ_{vci} were high in order for most of the daytime, although T_{ha} was higher than T_w. This indicates concurrent saline water evaporation and tubular cover inner surface condensation. Moreover, Figures 6(a) and (b) show that the evaporation and condensation periods during daytime are longer in September than December, indicating that the performance of the TSS is strongly related to the vapor movement within the tubular cover.

Assuming that the hourly production flux, m_{ph}, is in proportion to the evaporation flux from the saline water surface, $(\rho_{vw} - \rho_{vha})$ or the condensation flux on the tubular cover inner surface, $(\rho_{vha} - \rho_{vci})$, the following two equations could be proposed.

$$m_{ph} = a_e D_{atm} (\rho_{vw} - \rho_{vha}) \tag{1.2}$$

$$m_{ph} = a_c D_{atm} (\rho_{vha} - \rho_{vci}) \tag{1.3}$$

where, D_{atm} is the molecular diffusivity of water vapor in air, a_e is the evaporation coefficient and a_c is the condensation coefficient. The value of D_{atm} is a function of T_{ha} as follows (Miyazaki 1976):

$$D_{atm} = 209.52 \times 10^{-9} (T_{ha} + 273.15)^{2.3} \tag{1.4}$$

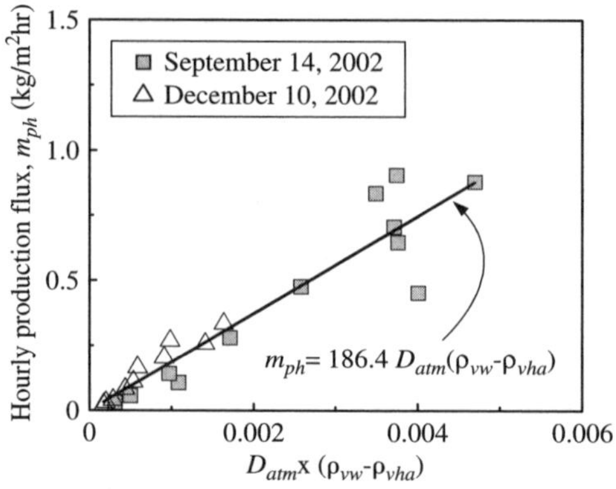

(a) Correlation between m_{ph} and evaporation flux

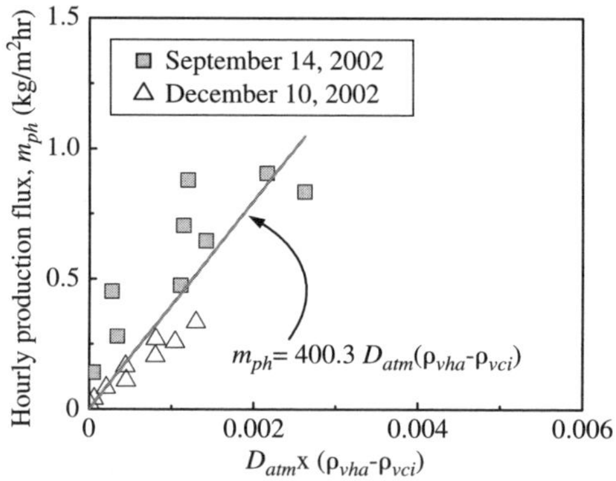

(b) Correlation between m_{ph} and condensation flux

Figure 7. Correlation between hourly production flux, m_{ph} and product of the molecular diffusivity of water vapor, D_{atm} and vapor density difference $(\rho_{vw} - \rho_{vha})$ or $(\rho_{vha} - \rho_{vci})$.

Figures 7 (a) and (b) show the relationship between m_{ph} and the product $D_{atm} \times \Delta\rho_{v1}$, and the relationship between m_{ph} and the product $D_{atm} \times \Delta\rho_{v2}$, respectively. The former implies the correlation between hourly production flux and evaporation flux, and the latter implies the correlation between hourly production flux and condensation flux. Consequently, a_e is $186.4\,\mathrm{m}^{-1}$ and a_c is $400.3\,\mathrm{m}^{-1}$. The value of m_{ph} increases with the increase in the $D_{atm} \times \Delta\rho_{v1}$ and the $D_{atm} \times \Delta\rho_{v2}$ products. Comparing m_{ph} to the same product of $D_{atm} \times \Delta\rho_{v1}$ and $D_{atm} \times \Delta\rho_{v2}$ in the morning and afternoon, it was slightly larger in the afternoon than the morning. The scattering of plots in Figure 7(a) is smaller than that in Figure 7(b). These results require further investigation on the production mechanism of the TSS taken account of the humid air physical properties.

1.3.4 *Conclusions*

A Tubular Solar Still (TSS) was proposed (designed by the authors) to rapidly and easily remove accumulated salt in the TSS, which have been the main drawbacks in the basin-type solar still. To provide better understanding of the mechanism in distilled water production, field experiments were carried out at the Hamuraniyah experimental farm in Ras Al Khaimah Emirate in the United Arab Emirates in 2002 and its logged data were analyzed.

The objective of the present study was to evaluate the effects of evaporation and condensation in the TSS on hourly production, calculated using the vapor density on the saline water surface, ρ_{vw}, the humid air vapor density, ρ_{vha} and the saturated vapor density on the tubular cover inner

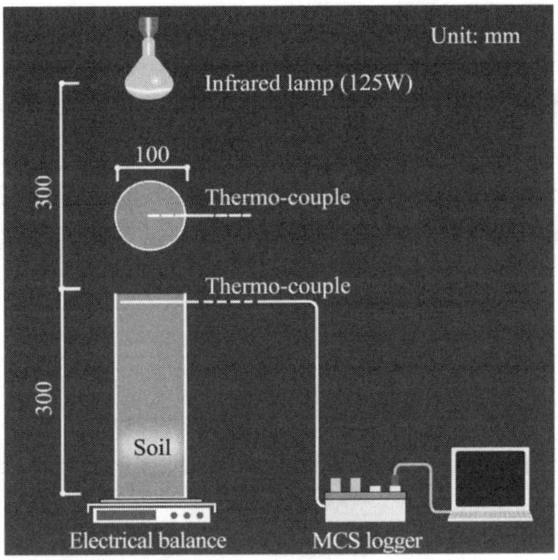

Figure 8. Indoor radiation evaporation experiment.

surface, ρ_{vci}. In order to accomplish the objective, simple bulk-type equations were proposed to calculate the evaporation flux from the saline water surface in the trough and the condensation flux generated on the tubular cover inner surface.

The following conclusions are drawn:

(1) Hourly production rapidly increases after sunrise and then descends gradually afternoon.
(2) The vapor densities ρ_{vw}, ρ_{vha} and ρ_{vci} are high in order for most of the daytime.
(3) The distilled water production is strongly affected by saline water surface evaporation and tubular cover inner surface condensation.
(4) The solar radiation is a prime factor that dominates the daily production of TSS.
(5) The present experiment will provide valuable theoretical production model taken account of the humid air physical properties.

2 IMPROVEMENT OF SOIL MOISTURE RETENTION BY DATE PALM MULCH

2.1 *Research purpose*

In order to actually reconcile economical agriculture and securement of water resource in the Middle East, the use of soil-moisture retention materials, which can be easily acquisitioned in the local area, will become the key of a success for the implementation of sustainable agriculture.

A date-palm tree represents a botanical resource in the Middle East, but may have not been frequently used in an agricultural field in this area and the surrounding area. In the present study, we prepared soil columns and date-palm chips as a mulch and conducted the heat and moisture transfer experiments in a constant temperature and humidity room to examine the performance of their mulch effects, i.e. evaporation control and suppression of soil-temperature rise associated with incoming short-wave (solar) radiation.

The purpose of the present study is to describe the performance of date-palm mulch effects, obtained from the heat and moisture transfer experiments.

2.2 *Indoor radiation evaporation experiment*

Dry Toyoura standard sand was packed in a column (height 0.3 m and diameter 0.1 m) with the bulk density of about 1600 kg/m^3. An infrared lamp (125 W) was emitted from a height of 0.3 m on the soil surface (See Figure 8). Three kinds of mulch columns, DP1 (mulch density = mass of mulch/

Table 1. Mulch column prepared for the experiment.

Non-treated	DP1	DP2	DP3	Water
Mulch density of date-palm (kg/m^2)				
0	0.25	0.7	1.0	–

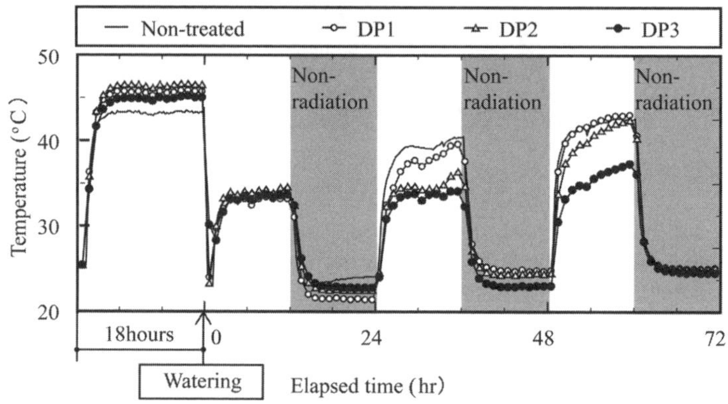

Figure 9. Time variation of soil surface temperature for different mulch columns.

mulching area: 0.25 kg/m^2), DP2 (0.7 kg/m^2), and DP3 (1.0 kg/m^2), and other two kinds of columns, i.e. a non-treated (bare soil) column and a water column were also prepared for the experiment (See Table 1). Starting radiating the lamp, the soil temperature was measured by a thermo-couple every 15 minutes for 18 hours, and then watering on the soil surface was performed for 8 minutes (the amount of watering: 12.7 kg/m^2). Lighting and turning off the lamp were repeated every 12 hours for three days after watering. During this experimental period, soil temperature was measured every 15 minutes, and each column weight was measured every 1 to 6 hours.

2.3 *Results and discussions*

2.3.1 *Soil temperature*

Figure 9 shows the time variation of soil surface temperature (1 cm below the soil surface). Although the mulch covered soil temperature was 2 to 3°C higher than the bare soil one before watering, all temperature levels suddenly fell down after watering, because of the sensible heat associated with watering and the latent heat of vaporization. As a result, the temperature difference was disappeared for the first radiation and non-radiation period. In the second radiation period, however, the soil temperature became high in order of the bare, non-treated and the reminder, i.e. DP1, DP2 and DP3. The bare soil temperature rose about 6°C, compared with that in the first radiation period, but the mulch covered soil temperature of DP1, DP2 and DP3 was the same as that in the first radiation period. This temperature difference between the bare soil and the mulch covered soil was attributed to the latent heat of vaporization acting on the mulch covered soil surface. In other words, the bare soil surface was not wet any longer then, while the mulch covered soil surface still maintained wet. In the third period, it suggests that the dryness of soil surface advanced further except the highest mulch density, DP3.

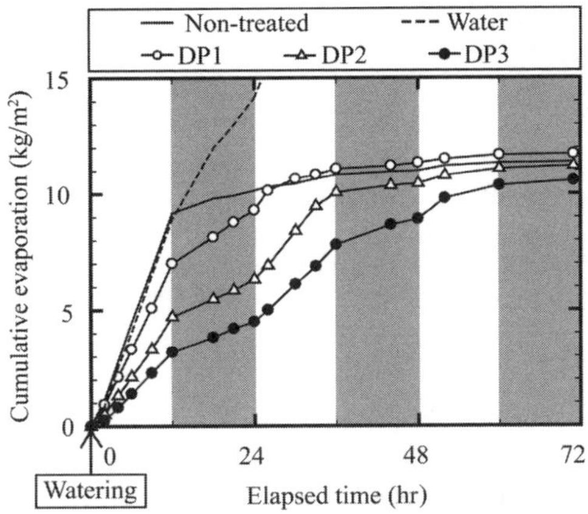

Figure 10. Time variation of cumulative evaporation for different mulch columns.

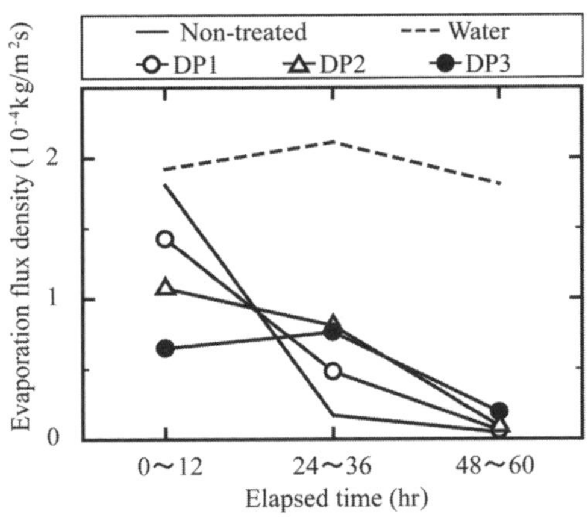

Figure 11. Time change of mean evaporation flux for different mulch columns.

2.3.2 *Evaporation control*

Figure 10 shows the time variation of cumulative evaporation per unit soil surface area. Comparing the cumulative evaporation from the bare soil with that from the mulch covered soil, the rate of evaporation reduction due to the date-palm mulch over the first radiation period was 24% for DR1, 49% for DP2 and 65% for DP3. As the mulch density increases, the evaporation control becomes clear. There was almost no difference in the cumulative evaporation between the bare soil surface and water surface. Of course, the evaporation flux, i.e. time rate of the cumulative evaporation is higher for the radiation period than the non-radiation period.

Figure 11 shows the time change of the mean evaporation flux during the radiation period. In the first radiation period, the evaporation flux becomes small in inverse proportion to the mulch density. The evaporation flux for the bare soil and water column are higher than that for the mulch covered soil column. In the second radiation period, the evaporation flux of the bare soil falls sharply and then gradually decreases in the third radiation period. This transition implies that the bare soil surface becomes desiccated. Comparing the evaporation flux of DP1 with DP2, the time rate of the diminution of the evaporation flux is higher for DP1 than DP2. The evaporation flux of DP3, however, keeps constant (steady evaporation stage) until the second radiation period. From this fact, it is implied that only the soil surface of DP3 was still wet until the end of the second radiation period. In the third radiation period, the transition from the constant-rate to the falling-rate evaporation stage appeared for DP3.

From the above results, drying and temperature rise of soil surface after watering become early with the decrease in the mulch density.

2.4 *Conclusions*

We have paid attention to date-palm chips as mulch from a viewpoint of the application of unused materials to agriculture in the Middle East. Heat and moisture transfer experiments using soil columns were, therefore, conducted in a constant temperature and humidity room to examine the evaporation control and sunshade effect by mulching the date-palm chips. Four kinds of mulch density (=mass of mulch/mulching area) were designed for the experiments, i.e. 0 (bare soil surface), 0.25, 0.7 and 1.0kg/m2, and the evaporation flux and soil temperature were measured.

The following conclusions can be drawn from the present study.

(1) The evaporation flux from soil surface decreases in proportion to the mulch density.
(2) Date-palm mulch can reduce the evaporation as high as 65% when the mulch density is 1.0 kg/m^2 and can lengthen a soil surface evaporation period.
(3) Consequently, the mulch covered soil can maintain low temperature for a long time rather than the bare soil.
(4) Drying and temperature rise of soil surface after watering become early with the decrease in the mulch density.

REFERENCES

Fukuhara, T., Fumio, A., Mutah, H.A., Nagai, N. & Yoshio, I. 1992. Production Mechanism and Performance of Tubular Solar Still. *Proceedings of the International Desalination Association*, Bahrain, BAH03-085.

Miyazaki, T.1976. Condensation and movement of water vapor in sand under temperature gradient. *Trans. JSIDRE*: 1–8.

Arid Land Hydrogeology: In Search of a Solution to a Threatened Resource – Mohamed(ed.)
© 2006 Taylor & Francis Group, London, ISBN 0 415 41127 0

Using pre-hydrated polymer to enhance soil moisture condition in arid environment

Y.Y. Aldakheel & F. Zeineldin
Water Studies Center, King Faisal University, Saudi Arabia

ABSTRACT: Agricultural polymers are known since the 1950's and their uses were limited because of the high prices. But, recent retail prices became more affordable due to extensive research and generated private sectors on this new technology. For instance, highly crossed – poly-acrylamide potassium based (HCPAM-K) hydrophilic gel, was found useful to improve water absorption capability and availability of sand dunes for medium saline water.

Most of cultivated farms at Al-Hassa oasis, Saudi Arabia depend on use of ground water as source for irrigation, where ground water is a limited resource and due to continuous pumping by huge quantities for irrigation and almost zero recharging, dramatic annual depletion is unavoidable. Introduce new techniques of water conservation to the agriculture industry, makes sense, save water and energy. One of recent promising technologies is use of artificial soil conditioners, such as polymers. In this study an experiment was conducted, to investigate effects of potassium based highly crossed polyacrylamide (HCPAM-K) hydro-gel polymer, on water holding capacity and percolation rate of three natural soils at Al-Hass oasis, using different irrigation water quality.

Preliminary results indicated that the amount of water adsorbed at 0.0 MPa pressure by sandy soil was increased by 33% with additions of 0.4% polymer. When the pressure was increased from 0 to 0.03 MPa, these additions enabled the sandy soil to retain more water up to 150%. This increase in water retention can reduce the amount of water otherwise lost by deep percolation. Moreover, the released amounts as compared to adsorbed soil water amounts at 0 MPa pressure, were 67%, 88% and 78% for clayey, sandy and polymer treated sandy soils respectively. This also indicated that sandy soil structure was improved and its adsorption ability increased by 10%, due to additions of polymer, manure and clayey soil.

1 INTRODUCTION

Al-Hassa oasis is located in eastern province of Saudi Arabia. It extends from latitude 25° 18′ 03″ N to latitude 25° 38′ 12″ N, and from longitude 49° 31′ 21″ E to longitude 49° 47′ 09″ E. The climate of the oasis in the summer (April to October) is severely dry and hot and no precipitation. Most of oasis' soils were consisted of sands and silts with some calcareous concretions. In general, sandy soils are known to be characterized by low water holding capacity and excessive percolation rates. In literature, amount of water leaking downward in sandy soil was estimated to be ranged between 864 and 8610 mm/d. Large amount of drainage waters from the oasis, is an eye witness to the affect of high permeability on deep percolation. To counteract irrigation water losses, farmers at Al-Hassa use natural soil conditioners such manure and clay for soil and water conservation.

Polymer related data and information were studied carefully, in order to choose right type of polymer to suit Al-Hassa saline soil and water conditions. One of the recent promising technologies is use of a hydro-gel polymer, which is known for its ability to improve saline water absorption capability and availability of sand dunes.

Sivapalan S. (2001) reported amount of water retained by a sandy soil at 0.01 MPa pressure was significantly ($p = 0.001$) increased by 23 and 95% with addition of 0.03 and 0.07% polymer, respectively. It was indicated the increase in water retention reduce the amount of water otherwise lost by deep percolation. When the pressure was increased from 0.01 to 1.5 MPa, the polymer enabled the sandy soil to retain more water. But the amount released from the soil was not significantly ($P = 0.95$) increased. Also, it was reported water use efficiency for plants increased by 12 and 19 with the addition of 0.03 and 0.07% polymer, respectively.

The water-holding capacity depends on the texture of the soil, the type of hydrogel and particle size (powder or granules), the salinity of the soil solution and the presence of ions. Cross-linked polyacrylamide hold up to 400 times their weight in water and release 95% of the water retained within the granule to growing plants. In general, a high degree of cross-linkage results in the material having a relatively low water-retention capacity. However, the water-holding capacity drops significantly at sites where the source of irrigation water contains high levels of dissolved salts (e.g., effluent water) or in the presence of fertilizer salts (Wang and Gregg 1990). It was indicated the amount of water retained by a soil, adversely affected by chemicals or ions (Mg^{2+}, Ca^{2+}, and Fe^{2+}) present in the water (Johnson, 1984b). James and Richards (1986) suggested that these divalent cations develop strong interactions with the polymer gels and are able to displace water molecules trapped within the polymer. Even though monovalent cations (Na^+) can also replace water molecules, the effect is not as pronounced as with the divalent counterparts as the process is fully reversible by repeated soaking with deionised water.

The effect of an amendment of sand soil with a highly cross-linked polyacrylamide (Stockosorb K 400) on the survival of *Pinus halepensis* seedlings during water stress was investigated (Huttermann et al., 1999). Varying concentrations of the hydrogel were added to sandy soils at 0.04%, 0.08%, 0.12%, 0.20% and 0.4% weight by weight. The highest addition, i.e., 0.4%, changed the water retention capacity and its change in water potential with regard to its water content from typical sand to a loam or even silty clay. Water retention of soil increased exponentially with increasing additions of Stockosorb K400 to sandy soil. During drought, the seedlings shown a pronounced growth both of the shoot and the roots, i.e., three-fold higher than plants in the control. It was indicated that amendment of soils with this type of hydrogel at 0.4% (w/w) will greatly enhance the drought tolerance of the seedlings growing on this substrate. The Stokosorb K is commercial name for potassium HCPAM-K hydro-gel polymer. It is highly cross-linked polyacrylamide with 40% of the amides hydrolysed to carboxylic groups.

A polyacrylamide hydrophilic gel (Agrosoak), was tested a soil conditioner for increasing the water absorbing capacity of sand dunes (Silberbush et al., 1993). It increased water availability and contributed to the increase in the yield of Cabbage (Brassica oleraceae L.) with saline water. Also it enabled replacement of an expensive drip irrigation system with an inexpensive sprinkling system. Different concentrations of the Agrosoak were added to the soil (0.00, 0.15, 0.3, and 0.45% by weight) at different water amounts and salinity levels (1.9–7.8 $dS.m^{-1}$). It was pointed Agrosoak may increase the water use efficiency with plants grown on sandy soils, when that crop is tolerant to Na^+.

Polyacrylamide is described as a superior water-absorbing synthetic polymer, i.e., hydro-gels, whose primary benefit is the dramatic increase in long-term, water holding capacity of soils (Daniel J. Wofford, Jr, and Dale Greenwood, 1991). When it is mixed into the soil, it absorbs and retains water via irrigation for extended periods, thus significantly increasing water availability to plants. A Fresno County, California, farmer flew on and incorporated 20 lbs. per acre (1989) on 25 sandy acres of a 160-acre canner tomato field. In previous years the farmer had never gotten full yields from the sandy field, but this year he could not differentiate between the sandy and non-sandy fields. Yields for both were in the 40 tons range; the sandy field had never before produced more than 27 tons per acre.

A hydrogel (Stockosorb K410) effects on growth and ion relationships of salt resistant woody species, *Populus euphratica*, were investigated under saline conditions (Chen et al., 2003). Addition of 0.6% hydrogel to saline soil, improved seedling growth (2.7 fold higher biomass) over a period of 2 years. Root length and surface area of treated plant had 3.5 fold higher than those grown in unamended saline soil. It was concluded that hydrogel treatment enhanced Ca^{+2} uptake and increased capacity of *P. euphratica* to exclude salt (i.e., reduces contact with Na^{1+} and Cl^{1}).

Therefore, improvement of sandy soil is an important task to conserve irrigation water and even winter rains in Al-Hassa region. The primary problem in improvement of sandy soil is to raise the water reserving ability of sandy soil to a level of usual arable land. Addition of HCPAM-K hydro-gel polymer may help to alleviate this problem and conserve irrigation water, by improving soils physical and chemical properties. With the reality of water restriction and scarcity facing growers in Al-Hassa oasis, as well as in Saudi Arabia, the benefit of using highly cross-link K-based poly-acrylamide as part of water conservation program should be thoroughly investigated. The objective of this research is to Evaluate water holding capacity and release ability of sandy soil that was treated with a mixture of hydro-gel polymer, manure and clay soil at different soil tensions ranged between (0.01 MPa) and (1.2 MPa) pressures.

2 MATERIALS AND METHODS

Three natural soils of irrigated farms were chosen, one was a lime farm at King Faisal university research station (F1), and other two were date palm farms at Al-Hassa irrigation project, i.e., one farm uses high saline drainage water, (F2) and the other uses mixed irrigation water (F3). Randomly, at each farm, three trees sites were chosen to represent replicates (i.e., R1, R2 and R3). Each replicate was subdivided into three equal sections, Treatment 1(T1), Treatment 2(T2) and Control(C1), see Figure 1.

Samples for determining soil physical and chemical properties of soils were taken from within fertilized zone of a tree (0–20 cm). Randomly, with an auger, 6 kg soil samples were sampled from each treatment site and the control, i.e., T1, T2 and C1. Samples were taken from 0–15 and 15–30 cm depth and placed in separate plastic bags, 3 kg each. Then soil samples of each replicate were put inside a 15 kg plastic bag labeled and brought to laboratory.

The following procedures were true for all soils and treatments. Soils were grounded, cleared from debris, sieved by 2 mm mesh, and placed into green house for drying. Similarly, dry sheep manure and clayey soil were prepared. In laboratory, pre-hydrated HCPAM-K formulation was prepared by addition of 12.5 grams of dry powder of HCPAM-K to every liter ($1000\,cm^3$) of irrigation water. This hydrated HCPAM-K was let to swell for 30–45 minutes and stirred several times, to assure complete absorption of irrigation water. In treatment 1 (T1), 1.21 liters of pre-hydrated HCPAM-K were added to 14 kg of soil, while 4.48 liters for treatment 2 (T2). Sandy soils were modified by addition of 10% manure and 10% clayey soil, while for soil with less sand such as loamy soil, fewer percentages were added. Then a modified soil was mixed with the pre-hydrated HCPAM-K formulation, to prepare pre-hydrated polymer treated (PPT) soil.

Standard packing procedure, which was given by (Klute et al., 1986), was employed to pack untreated soil into three infiltration cylinders (10.5 cm in dia, 50 cm in length). Initially, equal amount of untreated soil, i.e., obtained from 15 to 30 cm depth, were packed in the bottom of a cylinder,

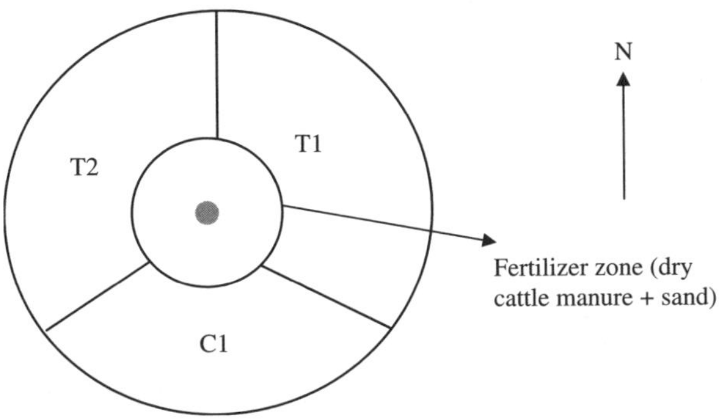

Figure 1. Layout of the treatments within the vicinity of a tree.

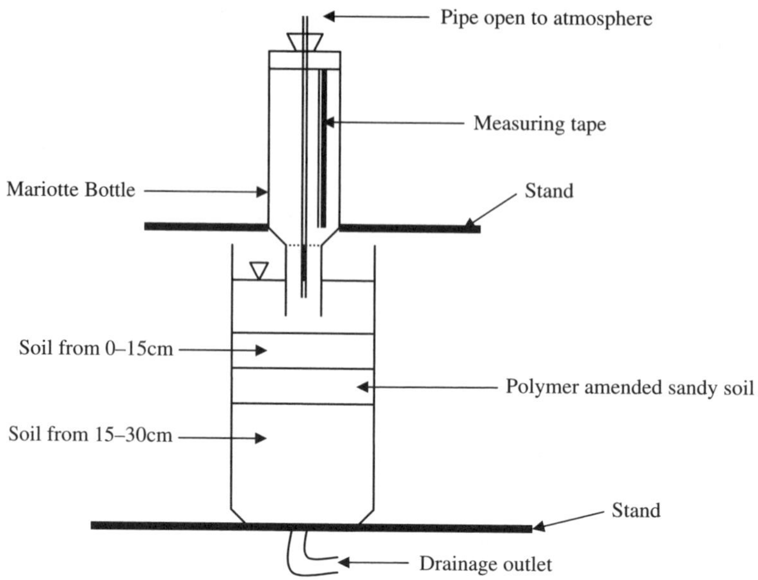

Figure 2. A diagram of an infiltration cylinder associated with Mariotte bottle.

up to 20 cm. Above that equal amounts of the PPT soil were packed, up to 10 cm in depth. Then surface layer of untreated soil, which obtained from 0 to 15 cm depth, was packed up to 10 cm (Figure 2).

Retain and availability of soil water experiment: Sandy soil (i.e., F1), clayey, and PPT soil were packed into standard cores (45.3 cm^3), in replicates of four, to a given field bulk density and weight. F1 soil was chosen, because it contained higher sand fraction than F2 and F3. They were then soaked into distilled water over night, to be saturated from the bottom, reweighed and placed into pressure plates. Retained and released soil water were assessed at water field capacity pressure (0.01 MPa) and wilting point pressure (1.2 Mpa).

3 RESULTS AND DISCUSSION

The amount of water adsorbed at 0.0 MPa pressure by sandy soil was increased by 33% with additions of 0.4% polymer, 10% sheep manure and 10% clay weight by weight (Figure 1). When the pressure was increased from 0 to 0.03 MPa, these additions enabled the sandy soil to retain more water up to 150%. This increase in water retention can reduce the amount of water otherwise lost by deep percolation. Released amounts as compared to adsorbed soil water amounts at 0 MPa pressure, were 67%, 88% and 78% for clayey, sandy and polymer treated sandy soils respectively. This also indicated sandy soil structure was improved and its adsorption ability increased by 10%, due to additions of polymer, manure and clayey soil. It was reported that the amount of water retained at relatively low values of matric suction (say, between 0 and 0.01 MPa of suction) depended primarily upon the capillary effect and the pore-size distribution, and hence was strongly affected by soil structure (Hillel, 1971).

In Figure 4, released amount of soil water against high tension (i.e., 0.065 to 1.2 MPa) when compared to 0.03 MPa in Figure 3, were more for amended sandy soil than clayey soil. That was because of fine pores of clayey soil role in dragging and resisting release of soil water, i.e., the release of soil water was affected by texture and specific area of clayey soil (i.e., 150–500 m^2/g). In all of the three soils, as (Figure 4), released soil water, at the high pressures, consistently increased with

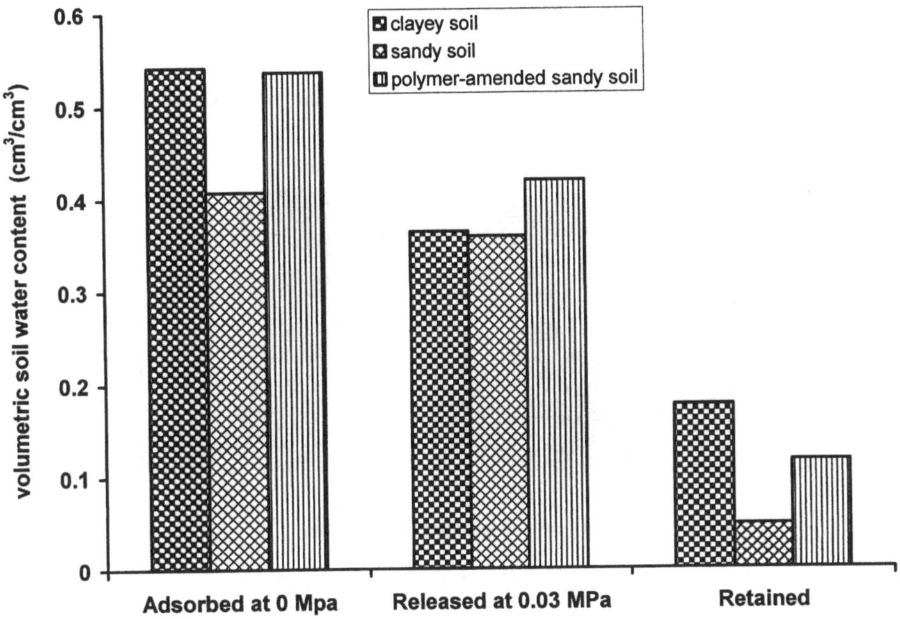

Figure 3. Adsorbed, released and retained soil water at 0 and at 0.03 MPa.

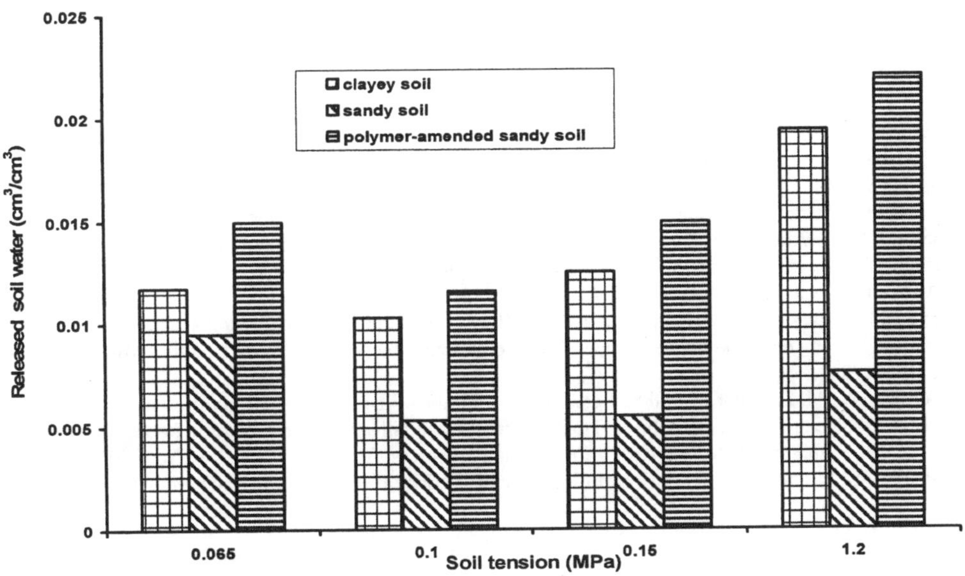

Figure 4. Released soil water against high soil tension.

the increase of soil tension from 0.1 to 12 MPa. Based on these results, more soil water is available to plants under polymer–manure-clay amended soil than sandy and clayey soil.

Soil moisture contents of sandy soil, against 0.03 and 1.2 MPa pressures, were increased by 144% and 170% respectively, when it was treated with 0.4% polymer, 10% manure, and 10% clayey soil (Figure 5). In fact, the difference of soil moisture between 0.03 and 1.2 MPA pressures, showed available soil water of amended soil was 125% more than untreated sandy soil. This indicated that

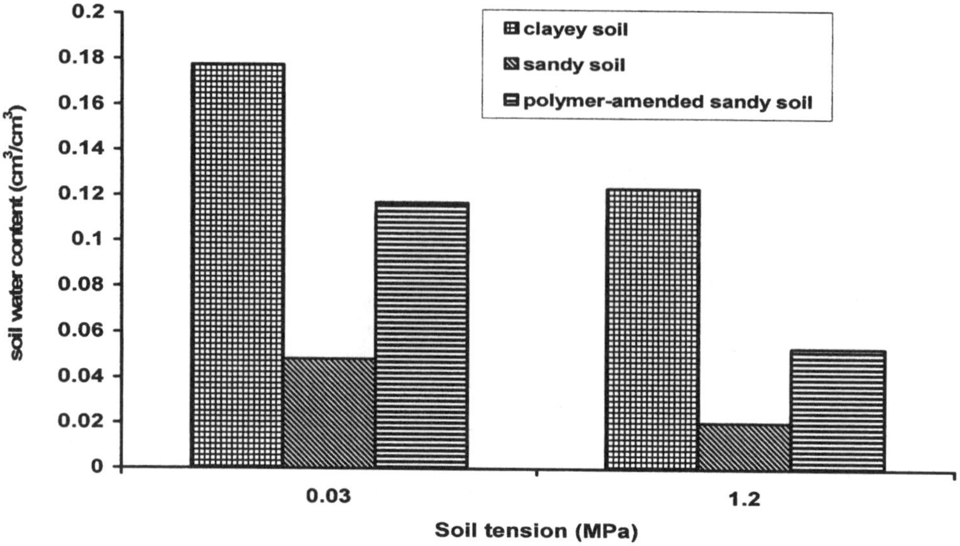

Figure 5. Soil moisture contents at 0.03 and 1.2 MPa tension and their difference.

amendment of sandy soil with polymer, manure and clayey soil had increased its specific surface area and therefore its adsorption rate to water. It was reported that water content at higher suction (1.5 MPa) was shown to be correlated with the surface area of a soil (Gardner, 1968).

REFERENCES

Gardner W.R. 1968. Availability and measurement of soil water. In "Water Deficits and plant growth," Vol. 1, pp. 107–135. Academic press. New York.

Sivapalan S. 2001. Effect of polymer on soil water holding capacity and plant water use efficiency. Proceedings of the 10th Australian Agronomy Conference, Australian Society of Agronomy.

Silberbush M., E. Adar and Y. De Malach. 1993. Use of an hydrophilic polymer to improve water storage and availability to crops grown in sand dunes II. Cabbage irrigated by sprinkling with different salinities. Agricultural water management. Vol. 23, Issue 4. 315–327.

Chen S., M. Zommorodi, E. Fritz, S. Wang and A. Huttermann. 2003. Hydrogel modified uptake of salt ions and calcium in *Populus euphratica* under saline conditions. Trees-Structure and Function. Vol. 18, Number 2. 175–183.

Huttermann A., M. Zommorodi and K. Reise. 1999. Addition of hydrogels to soil for prolonging the survival of Pinus halepensis seedlings subjected to drought. Soil and Tillage Research. Vol. 50, Issues 3–4. 295–304.

Hillel D. 1971. Soil and Water, physical principles and processes. Academic press. New York. 288 pages.

Kutilek M., D. R. Nielsen. 1994. Soil hydrology. Catena Verlag, 38162 Cremlingen-Destedt, Germany.

Arid Land Hydrogeology: In Search of a Solution to a Threatened Resource – Mohamed(ed.)
© 2006 Taylor & Francis Group, London, ISBN 0 415 41127 0

The use of treated oily-water for irrigation: Impacts on soil properties

M. Al-Haddabi & M. Ahmed
Department of Soils, Water and Agricultural Engineering, College of Agricultural and Marine Sciences, Sultan Qaboos University (SQU), Oman

ABSTRACT: Large quantities of water are produced during oil extraction in Oman. Disposal of this water poses serious challenge to the managers of the oil industry in Oman. The overall objective of this study was to assess the effect of treated oil production water on soil properties due to land disposal. A pilot scale study was conducted to treat the produced water continuously using a combined air floatation-adsorption process. A comparison study between the treated water and fresh water has been performed to test their effect on soil physical and chemical properties. Apart from increasing soil salinity, it has been found from the study that the use of treated water caused adverse effects on the physical properties of the soil such as pore size distribution, infiltration rate and saturated hydraulic conductivity. No significant difference was observed in heavy metals accumulation when fresh and treated water were used.

1 INTRODUCTION

During oil production in Oman, large quantities of water are also produced. Petroleum Development Oman (PDO) in 2002 produced some 330,000 m^3/day of water as a by-product of its oil output of approximately 135,000 m^3/day. The volume of this production water has been steadily increasing over time and was predicted to rise to about 650,000 m^3/day by the year 2005, as oil fields get older. Approximately 37% (123,000 m^3/day) of the currently produced water is utilized for reservoir pressure maintenance by injection mainly into northern Oman fields. About 140,000 m^3/day production water is presently disposed off into shallow (150–400 m) sub-surface formation at oil field sites mainly in south Oman (Al-Muscati et al. 1997). According to Oman environmental regulations, this practice should stop very soon. Presently, approximately 71,000 m^3/day is disposed of into deep sub-surface formations at oil fields in south Oman. The disposal of the produced water to deep formations is expensive. One of the possible alternatives to the underground disposal of produced water is land disposal by using it as irrigation water for salt tolerant crops. Special treatments are needed to clean up the produced water from its oil content. A pilot plant using an air-floatation and adsorption combined process has been operated successfully in treating oil contaminated water. The process successfully reduced oil in the water from 100–300 ppm to less than 0.5 ppm. A comparative study between treated oil production water and fresh water on soil physical and chemical properties has been performed.

2 MATERIALS AND METHODS

The objective of the soil experiment was to see how accumulation of salt and heavy metals are affected by different application rates for land disposal using vegetation. The field trials of the experiment lasted for two years with two different trials. In the first year test plots measuring 20 m × 20 m was used. The test plot was divided into six plots and three types of forage crops: Rhodes grass, Alfalfa, and Barley, were planted. While in the second year, plastic pots were used instead of the field plots used during the previous year. Experiments were conducted using both fresh and treated oily-water (Table 1). Soil samples were collected 4 times (in each field trial) during the experiment at monthly intervals. Soil samples were collected from 3 different depths with 4 replicates. The depths were 0–20 cm; 20–40 cm and 40–60 cm. Three application rates were used. Rate 1 application equals to potential evapotranspiration (ET_o) calculated using FAO CROPWAT model, whereas Rate 2 application was 1.5 times ET_o and Rate 3 application was 2.0 times ET_o. CROPWAT is a computer program for irrigation planning and management and was issued by Food and Agriculture Organization (FAO). CROPWAT calculates crop water requirements and irrigation requirements from climatic and crop data. Ministry of Agriculture and Fisheries, Oman uses this computer program for irrigation scheduling.

2.1 *Chemical analysis*

2.1.1 *Soil ECe and pH determination*

The soil extract was prepared by saturating the soil with distilled water and allowing it to stay overnight. On the next day the soil was subjected to a vacuum pressure (negative pressure) to extract the soil solution through filter paper, Whatman No.52. Then the extract was used for measurements. The ECe is measured by using Jenway 4020 Electrical conductivity meter. Soil pH is measured by using Jenway 3020 pH meter and Philips combine glass/colomel electrode type CE1 that had previously been calibrated at pH 7.0 and 9.0.

2.1.2 *Soil SAR determination*

For SAR determination, the extract from soil-saturated paste was used. Soluble cations of Na^+, Ca^{2+}, and Mg^{2+} were measured using Inductively Coupled Plasma (ICP-OES 3300). SAR for each soil was calculated by using the standard equation.

2.1.3 *Heavy metals analysis*

Soil samples were oven dried overnight at 105°C. Approximately one gram (1 g) of each sample was weighed into a 100 ml beaker. Twenty milliliters (20 ml) of nitric acid was added to each beaker. The beakers were covered and left to stand in a fume hood overnight. The beakers were placed on hotplates and warmed gently until frothing ceased and solutions were almost clear. At this stage the nitric acid was boiling gently without a major loss of volume (the beakers remained covered during this heating process). The beakers were allowed to cool on an asbestos mat. Three milliliters (3 ml) of perchloric acid was added to each beaker. The beakers were returned to the hotplates, without cover glasses, and gently heated to just dryness. The beakers were again cooled on an asbestos mat. Two milliliters (2 ml) of deionised water was added to each beaker, followed by two milliliters (2 ml) of 6 molar hydrochloric acid. The residues were dissolved by warming the beakers. The samples were then filtered and the solution was used for analysis by using the ICP.

2.2 *Physical analysis*

Soil columns were prepared, where soil was packed to 1.4 g/cm^3 bulk density. Experiments were conducted using both fresh and treated oily water. Basic water quality parameters are shown in Table 1. The columns were placed under the sun to achieve field conditions and to enhance the evaporation rate. Three water application rates were used for 1 month. The rates have been mentioned earlier.

Table 1. Water analysis.

Type	Na (mg/l)	Ca (mg/l)	Mg (mg/l)	EC (dS/m)	pH	SAR
Treated oily-water	900	34	20	3.62	7.8	30.28
Fresh water	100	9.25	7.5	0.56	7.6	7.8

2.2.1 *Saturated hydraulic conductivity (Ks)*

Ks was measured using "the constant head method", in which a constant level of water called head is kept above the soil surface. Then the water passing through the soil column is collected and measured in time intervals until it reaches constant velocity readings. To determine the Ks (cm/sec) the equation based on the Darcy's law was used.

2.2.2 *Infiltration rate*

Soil was packed in cylinders at 1.4 g/cm^3 bulk density. 1000 ml of water was applied to the soil sample in the cylinder and the water intake was recorded with time until the difference between water intake reached constant, then the infiltration experiment was terminated. The infiltration columns were left under the sun, irrigated with treated oily-water with three different rates for one month. The infiltration rates were calculated again after this one month.

2.2.3 *Pore size distribution*

Samples exposed to different application rates of treated oily-water were taken from the infiltration columns. Mercury porosimetry (Pore Master (Quantacrome)) was used in this experiment, where the volume of mercury penetrating the sample was measured as a function of the pressure imposed on the mercury. The "pore size" was calculated from this pressure by Weisbern's equation of capillarity and, using the "bundle of capillary tubes" model of pore structure, the volume of mercury is assigned to this "pore size".

3 RESULTS AND DISCUSSIONS

3.1 *Results of field trials*

3.1.1 *Soil salinity and sodicity*

It was found that the type of water used for irrigation is the major factor that has a significant effect in soil salinity. The average values of soil salinity measured as electrical conductivity of saturation extract (ECe) ranged between 2 and 7 dS/m for fresh water and between 2 and 22 dS/m for treated water (Figure 1). It is obvious that treated water causes a tremendous increase in soil salinity. This increase in salinity will lead to a point where the soil becomes an unsuitable place for growing crops. The addition of more water is thought to decrease the salts build-up through the process of leaching. However, from the statistical analysis of the results obtained we found that neither rate, nor depth or their interaction have significant effects on soil salinity. Another problem associated with application of the treated water was sodicity. Sodicity causes dispersion and swelling of clay minerals resulting in a structural breakdown and in a reduction in soil permeability. The use of treated water led to increase in the SAR of the soil (Table 2), which is a measure of sodicity. The main increase in sodicity occurs in the upper layer of the soil, where the SAR values increased for all the application rates.

3.1.2 *Soil pH*

The values of soil pH ranged between 7.6 and 8.5 for treated water and between 7.5 and 8.4 for the fresh water (Tables 3 and 4). It is clear that the type of water has no significant effect on pH. Both types of water have alkaline conditions. From the statistical results ($P < 0.05$), time and depth have significant effects on soil pH.

3.1.3 *Heavy metals*

At the beginning of this study, it was thought that the heavy metals along with salinity will be the main obstacle for using the treated water. The salinity can be overcome by using salt tolerant plants. The statistical analysis of the results shows that the treated oily-water has no significant effect on the accumulation of heavy metals. The high soil pH slows the solubility and mobility of the heavy metals in the soil. It was observed that the heavy metals accumulation is not a function of depth. Our objective was to investigate if there is any accumulation of heavy metals when using treated water when it is compared with fresh water but not to look in the chemical behavior of each

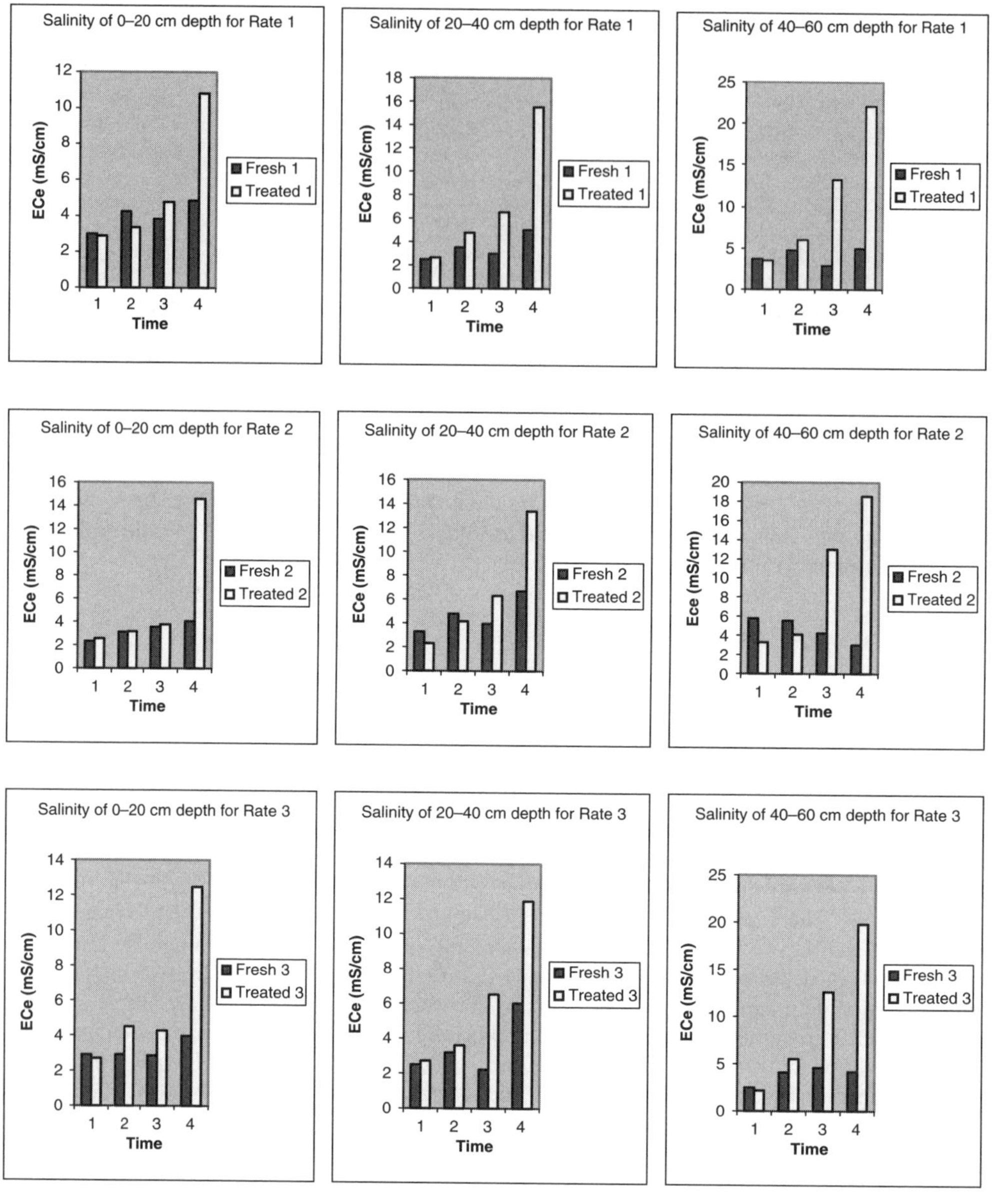

Figure 1. Soil salinity data (ECe in dS/m) for different application rates at different depths.

element in the soil, which is beyond the scope of this study. Comparing analytical data with the Omani standards for the use of wastewater in irrigation, it is concluded that the treated water fulfills the minimum standard for irrigation water with respect to heavy metals. The results show also that there is no significant difference between fresh water and treated water with regards to accumulation of heavy metals (Table 5).

3.2 *Soil physical properties*

Al-Haddabi et al. (2004) provides information on the changes in soil physical properties that took place due to application of treated oil production water. The infiltration rate (f) was calculated by

Table 2. SAR of soil receiving water with different application rates.

Type of water	Time	Depth (cm)	Rate 1	Rate 2	Rate 3
Fresh water	1	0–20	3.55	4.3	4.75
		20–40	5.55	5.3	5.75
		40–60	4.3	5.05	5.2
	2	0–20	3.45	3.85	4.75
		20–40	4.1	4.5	5.35
		40–60	3.85	5.5	5.0
	3	0–20	3.9	3.3	11.35
		20–40	3.7	4.55	6.3
		40–60	3.3	3.8	4.1
Treated water	1	0–20	4.3	5.55	5.35
		20–40	6.3	5.5	7.45
		40–60	4.2	5.15	6.45
	2	0–20	15	16.35	22.7
		20–40	6.15	6.7	11.5
		40–60	9.6	5.6	8.75
	3	0–20	6.75	16.1	11.75
		20–40	7.7	12.45	5.65
		40–60	3.3	4.4	4.15

Table 3. pH of soil applied with treated water.

	Depth (cm)		
Time	0–20	20–40	40–60
1	7.9	7.9	7.8
2	7.7	7.7	7.8
3	7.9	7.8	8.0
4	8.0	8.0	8.2

Table 4. pH of soil applied with fresh water.

	Depth (cm)		
Time	0–20	20–40	40–60
1	7.9	7.9	7.9
2	7.7	7.7	7.8
3	7.8	7.8	8.0
4	7.9	8.0	8.2

using the Kostiakov equation, $F = A\,t^B$ where F is the cumulative infiltration depth (cm) over the time t (minutes), A and B are empiric infiltration parameters that depend upon soil properties and initial conditions. The data in Figure 2 reveal that f (infiltration rate) tends to decrease with time and also with increasing application rates. The application rates here refer to the amount of water applied daily for a month prior to conducting the infiltration experiment.

The reduction in the infiltration rate can be related to sodicity, which tends to decrease the infiltration rate through dispersion and deposition of clay particles in the narrow necks of the conducting pores affecting the pore size distribution. The reduction in the infiltration rates could be different than the values obtained because the soil conditions were not exactly the same when infiltration experiments were conducted. When the initial infiltration experiments were conducted

Table 5. Copper, Lead & Zinc in the soil extracts (in ppm) of the whole profile (0–60 cm).

		Treated water irrigation			Fresh water irrigation		
Analyte	Time	Rate 1	Rate 2	Rate 3	Rate 1	Rate 2	Rate 3
Copper	1	6.96	6.46	5.55	6.00	6.47	6.53
	2	5.1	5.14	5.04	5.97	5.76	5.28
	3	5.75	5.57	5.15	5.44	4.80	5.05
Lead	1	3.52	2.57	2.81	3.01	2.96	3.0
	2	4.27	2.81	2.80	2.71	3.01	3.07
	3	2.05	1.89	1.69	1.21	1.23	1.85
Zinc	1	6.35	4.32	4.04	3.71	4.76	4.67
	2	3.3	2.52	2.78	2.75	4.11	2.88
	3	5.32	4.73	4.45	3.93	3.62	3.79

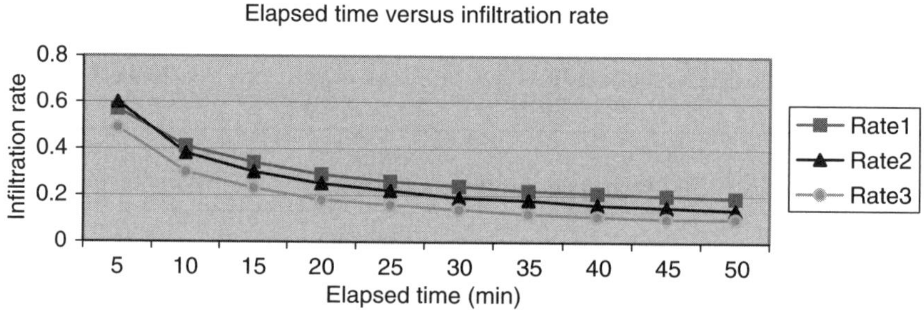

Figure 2. Infiltration rates in cm/min (after 1 month irrigation) calculated using fitted Kostiakov equation.

Table 6. Pore volume in soil samples (in percentage).

Type of water/rate	0–20 μm	20–50 μm	50–100 μm	100–200 μm	>200 μm
Fresh	0	41	39	18	2
Treated (Rate 1)	36	28	20	16	0
Treated (Rate 2)	35	33	21	11	0
Treated (Rate 3)	27	35	22	15	1

the soil was totally dry but after one month of irrigation the moisture content was increased. The soil columns were left to dry under the sun for several weeks. But the moisture content was still higher than the initial conditions.

A reduction in the saturated hydraulic conductivity occurred when oil-treated water was used. The reduction reached up to 43% of the initial value. This is due to the presence of high sodium levels. Hydraulic conductivity and permeability tend to decrease as Exchangeable Sodium Percentage (ESP) increases and salt concentration decreases (Quirk and Schofield 1955; McNeal and Coleman 1966; McNeal et al. 1968). Quirk and Schofield (1955) postulated that this is caused by clay swelling, clay dispersion and the subsequent plugging of conducting pores by the dispersed clay, and failure of soil aggregates. Data from pore size analysis show (Table 6) the volume of pores with a diameter 200–20 μm has been reduced while the volume of pores with a diameter <20 μm is increased when treated water is used. The change in the volume of pores could be due to the dispersion of soil particles, which narrows the conducting pores.

4 CONCLUSION

The application of treated-oily water even for a short duration on land has an adverse affect on chemical and physical properties of the soil. Due to the high salt content of the treated oily-water, it is found to have a significant effect on soil salinity. The results demonstrate that neither water application rates, nor depth or their interaction have significant effects on soil salinity. The rate of water application was thought to have significant effects on soil salinity but results prove no such relationship. Statistically, treated water had no significant effects on soil pH and heavy metals accumulation. High sodicity of treated oily-water causes reductions in infiltration rate and saturated hydraulic conductivity. The volume of large pores was reduced whereas the volume of small pores increased when treated oily-water was used. These results suggest that disposal of treated oily-water through irrigation and land spreading will be difficult. Impact on soil properties after long-term exposure to such waters needs to be investigated.

REFERENCES

Al-Muscati, A., Huijskes, J., and Parker, D.H. 1997. Production water management in Oman. Proceedings of Middle East Oil Show. Bahrain. 265–271.

Al-Haddabi, M., Ahmed, M., Kacimov, A., Rahman, S., and Al-Rawahy, S. 2004. Impact of treated wastewater from Oil Extraction Process on Soil Physical Properties. Communications in soil science and plant analysis, vol 35 (5 & 6).

McNeal, B.L., and Coleman, N.T. 1966. Effect of solution composition on soil hydraulic conductivity. Soil Sci. Soc. Am. Proc. 30: 308–312.

McNeal, B.L. 1968. Prediction of the effect of mixed-salt solutions on soil hydraulic conductivity. Soil Sci. Soc. Am. Proc. 32: 190–193.

Quirk, J.P., and Schofield, R.K. 1955. The effect of electrolyte concentration on soil permeability. J. Soil Sci. 6: 163–178.

Arid Land Hydrogeology: In Search of a Solution to a Threatened Resource - Mohamed(ed.)
© 2006 Taylor & Francis Group, London, ISBN 0 415 41127 0

Effects of polyethylene colors and thickness on the efficiency of soil solarization under the environment of the UAE

A. El-Keblawy, T. Ksiksi & F. AL-Hammadi
Department of Biology, Faculty of Science, UAE University, Al Ain, UAE

ABSTRACT: The effects of soil mulching with different polyethylene colors and thicknesses on seed viability of *Portulaca oleracea* (common purslane) were evaluated through the estimation of final germination percentage and germination rate. Common purslane is often used as an indicator species such that its control with solarization will often result in control of most weed species. This experiment was carried out from June 10 to July 25, 2003 in a farm near Al-Ain, UAE. Fresh seeds of *P. oleracea* were buried in the soil at 2.5 cm and 15 cm depths for 15, 30 and 45 days. The used soil polyethylene colors and thickness were: 120 μm opaque black, 50 μm transparent (thin), 75 μm transparent (medium) and 150 μm transparent (thick). The exhumed seeds were germinated in an incubator adjusted to 15/25°C. The results showed that the most effective plastic (polyethylene) sheets in deteriorating seed germination were the thin and medium thickness of the transparent sheets. The deterioration in seed viability increased with the increase in solarization duration and this was more pronounced in thin transparent plastic than in thick transparent and black plastics. The deteriorative effect of solarization on germination speed was less than that on final germination. The germination was significantly slower at 2.5 cm depth after 45 days solarization under all plastic types and at 2.5 and 15 cm after 30 and 45 days solarization under the transparent thin plastic. This further supports the efficiency of thin transparent in deteriorating the seeds of the two depths after shorter period of time. The results strongly recommend the use of thin or medium thickness of transparent sheets instead of black sheets, which are regularly used in the UAE.

1 INTRODUCTION

The substantial expansion in agricultural production in United Arab Emirates (UAE) in recent years has been associated with development and spread of several weeds and other soil-borne pathogens. Methyl bromide, the sodium salt of metham and dazomet is widely used in the United Arab Emirates for soil fumigation in open fields and in green houses (Al-Masoum et al. 1993). However, these fumigants are considered to be harmful for all organisms. The loss of methyl bromide fumigant to the air and possibly reaching groundwater as a contaminant and its potential ozone reduction has stimulated interest in finding an alternative method for disinfestation of soil–borne pests and weeds. Soil solarization was considered as appropriate, non chemical, alternative to methyl bromide fumigation.

Soil solarization is a natural, hydrothermal process of disinfesting soil of plant pests by using passive solar heating of moist soil mulched with polyethylene sheets (Katan 1981). This simple technique has the potentiality to increase the crop yield (Al-Masoum et al. 1993, 1998, Chellemi et al. 1999), improve soil chemical characters (Arora and Yaduraju 1998, Ahmad et al. 1996). In addition, this environmentally friend technique does not have negative effects on many of beneficial microorganisms such as arbuscular mycorrhizal fungi (Schreiner et al. 2001) and nitrogen fixing bacteria (Ricci et al. 1999).

The effectiveness of soil solarization for direct thermal inactivation of soil borne-pests is dependant on a number of physical factors including solar radiation intensity and duration, air temperature, amount of humidity at the soil surface beneath the tarps, properties of plastics used to produce the green house effect resulting in the heating of soil and properties of the soil to be treated (Stapleton and Devay 1986, Stapleton 2000). The UAE characterizes by higher air temperature and intensive solar radiation for longer duration, especially during summer months. Soil solarization showed great efficiency under the environment of the UAE (Al-Masoum et al. 1993, 1998, El-Keblawy et al. 2004). The recorded temperatures in a solarization experiment during the summer under the environment of the UAE were much higher than that recorded in most solarization studies in many parts of the world (El-Keblawy et al. 2004). Maximum temperature ranged between 33.5°C and 70.1°C at soil surface and between 40.1°C and 52.6°C at 10 cm depth. The maximum increases in soil temperature over that of control were 31.5°C at soil surface and 16.7°C at 10 cm depth (El-Keblawy et al. 2004). Similarly, Al-Masoum et al. (1998) in a field experiment conducted in Al-Ain, UAE found that the maximum temperatures reached under the transparent cover were 60.3°C, 53.3°C and 51°C at depths 5, 10, and 20 cm, respectively. On the other hand, the maximum increases in soil temperature by solarization over that of control at 10 cm depth were 16.5°C in Iraq (Brighton 1972) and 11.5°C at 10 cm in Pakistan (Ahmad et al. 1996). At 5 and 15 cm, the increases were 8.9°C and 11.5°C, respectively at south Italy (Mauromicale et al. 2001). In India, the increase in soil temperature over that of control was 8.1–9.5°C (Mudalagiriyappa et al. 1999).

It has been demonstrated that transparent plastics promote a relatively large net radiation at the soil surface, increase soil heat flux and, as a consequence, the minimum and maximum soil temperature are increased. However, opaque black and white plastics decrease the soil heat flux and the daily amplitude of the soil temperature (Devay 1991, Streck et al. 1995). In several countries, transparent polyethylene sheets was more effective in increasing soil temperatures and consequently in controlling soil pathogens and weeds than black polyethylene sheets (e.g., Iraq, Brighton 1972; India, Mudalagiriyappa et al. 1999; UAE, Al-Masoum et al. 1998, El-Keblawy 2002, El-Keblawy et al. 2004). In the UAE, however, most farmers commonly use black, instead of transparent polyethylene for soil disinfestations.

Portulaca oleracea L. (common purslane) appears to be cosmopolitan in distribution, especially in tropics, and is reported to be a summer annual outside the tropics (see Singh 1973). It is a successful colonizing species and is recorded to be one of the eight most common weeds in the world (Zimmerman 1976, Holm et al. 1977). *P. oleracea* is one of the annual species cited in association with solarization experiments. It is often used as an indicator species such that its control with solarization will often result in control of most weed species (Egley 1983, 1990, Duranti and Cuocolo 1988, Abu-Irmaileh 1991, El-Keblawy et al. 2004). El-Keblawy et al. (2004) studied the effect of soil solarization on seed viability of three annual weeds under the environment of the UAE and found *P. oleracea* was the most resistant weed. The aim of this study was to assess the efficiency of soil mulching with different polyethylene colors and thickness on final germination percentage and germination rate of *Portulaca oleracea* seeds. The efficiency of the plastic types on seed germination will be tested for seeds buried at two depths (2.5 and 15 cm) for different solarization durations (15, 30 and 45 days).

2 MATERIALS AND METHODS

2.1 *Seed collection and solarization experiment*

The experiment was conducted during the hottest period of the year from June 10 to July 25, 2003 in a private farm near Al-Ain, UAE (Latitude N24° 44, Longitude E55° 46 and Altitude 306 m above sea level). The farm soil was sandy loam. Seeds of *Portulaca oleracea* were collected from different farms around Al-Ain city in early June 2003. Seeds were cleaned and placed into 4 cm × 6 cm mesh bags. The mesh size of the bags was small enough to keep the seeds and allow free movement of water, gases exchange and microorganisms.

Prior to solarization, the soil was plowed, leveled and divided into 24 individual plots, 20 × 2 m each. Drip irrigation lines were laid in the middle of the plots. The soil was heavily irrigated in early June 2003. The experiment was conducted in a randomized complete block design with two replications and three factors namely are 4 types of polyethylene plastic mulches (120 μm opaque black, 50 μm transparent, 75 μm transparent and 150 μm transparent), two seed burial depth (2.5 cm and 15 cm) and three solarization durations (15, 30 and 45 days). In each plot, seeds were buried at the two depths. In order to keep high soil moisture content throughout the experiment soil was watered every week with the drip irrigation system.

2.2 *The germination experiment*

After 15, 30 or 45 days of soil solarization, seeds of *P. oleracea* were exhumed from 2.5 cm and 15 cm depths of 8 plots (two for each plastic type). The collected seeds were air dried and the seeds of each two replicates of the same treatment were mixed together. Mixed seeds for each treatment were tested for viability by germination in an incubator set at 15/25°C with dark condition coincided with 15°C. Some of the fresh seeds of *P. oleracea* were tested for germination as control without being subjected to solarization. The germination test was conducted in 9 mm plastic Petri-dishes with one Whatman No. 1 filter paper moistened with distilled water. Four replicate dishes were used for each treatment. Radical emergence was the criterion for germination to be recorded. Germinated seedlings were counted and removed every alternative day for 20 days post-seed sowing

2.3 *Calculations and statistical analysis*

The rate of germination was estimated using a modified Timson index of germination velocity $= \Sigma G/t$, where G is the percentage of seed germination at 2 days intervals and t is the total germination period (Khan and Ungar, 1984). The maximum value possible here using this index was $1000/20 = 50$. The greater the value, the more rapid is the germination.

Three-way analyses of variance (ANOVAs) were carried out to test effects of main factors (type of plastic, depths of seed burial and solarization duration) and their interactions on the final germination percentage and germination rate. T-tests were used to compare both final germination and germination rate of *P. oleracea* between the control (non-solarized seeds) and seeds solarized at the different treatments (plastic type, burial depth and durations). Tukey test (Honestly significant differences, HSD) was used to estimate least significant range between means. The germination percentages were arcsine transformed to meet the assumptions of ANOVA. The transformation improved normality of distribution of data. All statistical methods were performed using SYSTAT, version 11.0.

3 RESULTS

3.1 *Effects on final germination*

Three-way ANOVA showed significant effects for the main factors (type of plastic, depths of seed burial and solarization duration) on final germination percentage of *P. oleracea* seeds ($P < 0.001$, Table 1). The most effective plastic sheets in deteriorating seed germination were the thin and

Table 1. Three-way ANOVA showing the effects of depths of seed burial, solarization duration and type of plastic on the final germination percentage of fresh seeds of *Portulaca oleracea*. NS = insignificant different at P = 0.05.

Source of variation	df	Mean Square	F Ratio	P
Plastic type (P)	3	0.272	29.12	<0.001
Duration (Du)	2	0.755	80.81	<0.001
Depth (D)	1	0.600	64.25	<0.001
P × Du	6	0.030	3.33	<0.05
P × D	3	0.013	1.41	ns
Du × D	2	0.027	2.94	ns
P × Du × D	6	0.005	0.535	ns
Error	72	0.009		

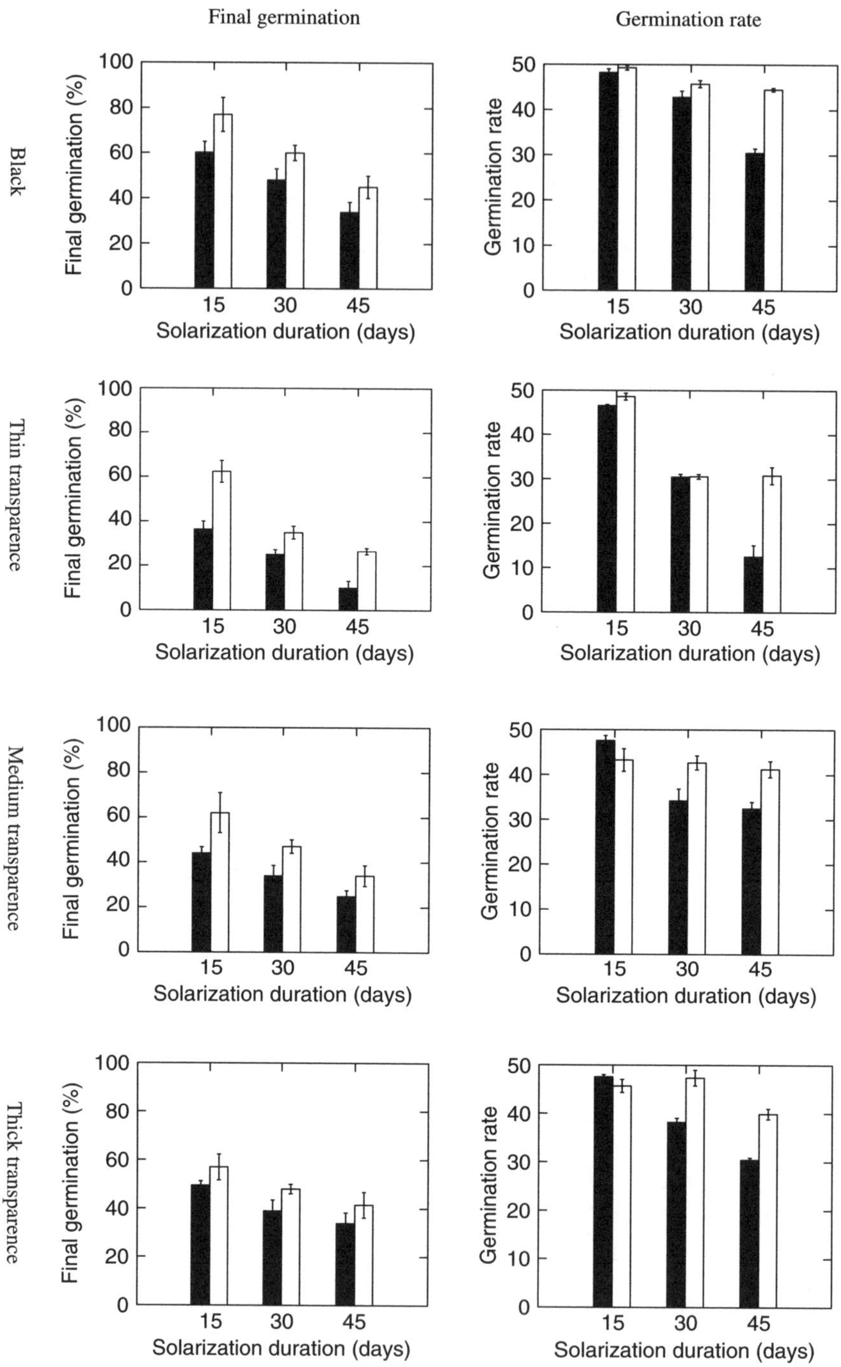

Figure 1. Effects of polyethylene thickness on final germination percentage and germination rate of *Portulaca oleracea* seeds solarized at different depths of the soil for different durations. Black bars = 2.5 cm and white = 15 cm depths. Germination percentage and rate for control were 80% ± 5.4 and 44.9 ± 0.8, respectively.

Table 2. Effects of plastic color and thickness, depth of seed burial and solarization duration on deterioration in final germination percentages (means ± SE) of fresh seeds of *Portulaca oleracea.*

Polyethylene type	Depth	Duration (days) 15	30	45	Overall
Black	2.5	−24.7 ± 5.2	−39.9 ± 5.1	57.6 ± 5.1	−43.8 ± 4.6
	15	−3.7 ± 7.8	−25.0 ± 3.5	−43.7 ± 5.2	−24.2 ± 5.8
	Overall	−15.4 ± 5.6	−32.5 ± 4.0	−52.9 ± 4.2	−35.7 ± 4.0
Transparent thin	2.5	−54.5 ± 3.8	−68.8 ± 2.4	−87.5 ± 3.2	−70.2 ± 4.4
	15	−22.0 ± 5.1	−56.2 ± 2.9	−66.8 ± 1.5	−48.4 ± 6.1
	Overall	−38.2 ± 6.8	−62.5 ± 2.9	−77.1 ± 4.2	−59.3 ± 4.3
Transparent medium	2.5	−45.0 ± 2.9	−57.5 ± 4.8	−68.9 ± 3.1	−61.2 ± 3.1
	15	−22.5 ± 8.5	−41.3 ± 3.1	−57.5 ± 4.8	−42.0 ± 5.1
	Overall	−35.3 ± 5.8	−49.3 ± 4.0	−65.7 ± 2.8	−53.8 ± 3.2
Transparent thick	2.5	−38.1 ± 2.1	−51.2 ± 4.7	−57.6 ± 5.1	−48.5 ± 3.9
	15	−28.8 ± 5.5	−40.0 ± 2.0	−48.1 ± 5.5	−38.9 ± 3.4
	Overall	−35.0 ± 2.5	−45.6 ± 3.1	−54.4 ± 3.9	−44.9 ± 2.4

medium thickness of the transparent type. Tukey test showed that the final germination of both of these plastics (32.6% and 36.9%, respectively) was significantly lower than that of thick transparent and black sheets (44% and 51.5%, respectively). Deterioration in seed germination increased significantly with the increase in solarization duration. The overall final germination decreased from 55% after 15 days to 42% and 31% after 30 and 45 days, respectively. Similarly, final germination was significantly greater at 15 cm depth (49.3%) than 2.5 cm depth (36.3%, Fig. 1).

The effect of the interaction between plastic type and duration was significant ($P < 0.05$, Table 1). The decrease in the final germination with the increase in solarization duration was more pronounced for seeds of thin transparent than those of thick transparent and black sheets. The final germinations for seeds solarized under transparent thin, transparent thick, and black sheets were 49.4%, 52% and 67.7%, respectively after 15 days and were 18.3%, 36.4% and 37.6%, respectively after 45 days. The deteriorations in the seed germination after 15 days, compared to non-solarized seeds (control), were 38.2%, 35.3% and 15.4% for transparent thin, transparent thick and black plastics, respectively. After 45 days, the deteriorations were 66.8%, 54.4% and 52.9%, respectively (Table 2, Fig. 1). The results showed a considerable proportion of *P. oleracea* seeds resisted solarization for 45 days at 15 cm depth, so this proportion depended on plastic types and thickness. It varied between 26% under thin transparent sheets to 34%, 36.4% and 45% under medium transparent, thick transparent and black sheets (Fig. 1).

3.2 *Effects on timson rate*

Three-way ANOVA showed significant effects on germination rate of *P. oleracea* seeds for the main factors (type of plastic used in solarization, depth of seed burial and solarization duration, $P < 0.001$, Table 3). Generally the deteriorative effect of solarization on germination rate was less than that on final germination (Fig. 1 and Table 4). Tukey test showed that the germination rate of seeds solarized under thin transparent sheets was significantly lower than that of seeds of the other plastic types. The effects of duration and depth of seeds burial on germination rate was similar to their effects on final germination percentage (Fig. 1).

The effect of the interaction between each pairs of the main factors and the three of them on germination rate was significant ($P < 0.001$). The exception was the interaction between plastic type and depth of seed burial ($P > 0.05$, Table 2). The germination rate of seeds buried at both 2.5 cm and 15 cm depth after 15 days of solarization was greater than that of the control (44.9), so the difference was insignificant. Also, germination of seeds at 15 cm depth was faster than or equal that of the control under the black plastic after 30 and 45 days and under thick transparent plastic after 30 days of solarization. The significant reductions in germination rate were at 2.5 cm depth after

Table 3. Three-way ANOVA showing the effects of depths of seed burial, solarization duration and type of plastic on germination rate of fresh seeds of *Portulaca oleracea*. NS = insignificant different at P = 0.05.

Source of variation	df	Mean square	F ratio	P
Plastic type (P)	3	496.8	81.7	<0.001
Duration (Du)	2	1869.9	307.7	<0.001
Depth (D)	1	824.2	135.6	<0.001
P × Du	6	151.9	25.0	<0.001
P × D	3	6.9	1.13	ns
Du × D	2	415.6	68.4	<0.001
P × Du × D	6	44.8	7.38	<0.001
Error	72	6.08		

Table 4. Effects of plastic color and thickness, depth of seed burial and solarization duration on deterioration in germination rate (means ± SE) of fresh seeds of *Portulaca oleracea*.

Polyethylene Type	Depth	Duration (days) 15	30	45	Overall
Black	2.5	7.5 ± 1.6	−4.6 ± 2.5	−32.0 ± 1.9	−13.9 ± 4.5
	15	9.8 ± 0.9	1.9 ± 1.4	−0.8 ± 0.7	3.6 ± 1.5
	Overall	8.6 ± 1.0	−1.3 ± 1.8	−21.6 ± 4.6	−6.7 ± 3.2
Transparent thin	2.5	3.7 ± 1.4	−32.0 ± 1.1	−72.1 ± 4.9	−33.5 ± 9.5
	15	8.3 ± 0.9	−31.9 ± 0.9	−31.4 ± 3.6	−18.3 ± 5.8
	Overall	6.0 ± 1.1	−32.0 ± 0.7	−51.7 ± 8.2	−25.9 ± 5.6
Transparent medium	2.5	6.1 ± 4.2	−23.7 ± 4.8	−27.6 ± 3.1	−19.2 ± 3.8
	15	−3.6 ± 2.8	−4.9 ± 2.9	−8.0 ± 3.3	−5.7 ± 1.9
	Overall	1.9 ± 2.7	−14.3 ± 4.4	−21.9 ± 3.4	−14.1 ± 2.7
Transparent thick	2.5	6.0 ± 0.9	−14.9 ± 1.6	−32.2 ± 0.9	−13.4 ± 3.9
	15	1.9 ± 2.5	5.6 ± 2.7	−11.1 ± 2.0	−1.8 ± 2.6
	Overall	4.7 ± 1.2	−6.1 ± 4.3	−25.2 ± 3.1	−9.3 ± 2.9

45 days solarization under all plastic types and at 2.5 and 15 cm after 30 and 45 days under the transparent thin plastic (Fig. 1 and Table 4). These results further support the greater efficiency of thin transparent, compared to the other sheets.

4 DISCUSSION

The results of the present study showed that the transparent polyethylene sheets were more effective in the deterioration of *Portulaca oleracea* seed germination than black sheets. In addition, thin (50 μm) and medium (75 μm) transparent sheets were more efficient than thick transparent sheets (150 μm). Several studies in different regions of the world showed that transparent polyethylene is more effective in increasing soil temperatures and consequently in controlling soil pathogens and weeds, than black polyethylene. For example, in Iraq, soil temperature at 10 cm depth was greater under transparent sheets (49.7°C) compared to under black polyethylene sheets (45.9°C) (Brighton 1972). Similarly, the increase in soil temperature by solarization in India, over that of control, was 8.1–9.5°C for transparent sheets but was 2–3.4°C black sheets (Mudalagiriyappa et al. 1999). In Al-Ain, UAE, the maximum temperatures reached under the transparent sheets were 60.3°C, 53.3°C and 51°C at depths 5, 10, and 20 cm, respectively (Al-Masoum et al. 1998) and ranged between 33.5°C and 70.1°C at soil surface and between 40.1°C and 52.6°C at 10 cm depth (El-Keblawy et al. 2004).

Polyethylene is an ideal film for solar heating of soil because it is essentially transparent to solar radiation (280–2500 nm), extending to the far infra-red, but much less transparent to terrestrial radiation (5000–3500 nm), and thus reducing the escape of heat from the soil (Devay 1991). On the other hand, black polyethylene contains carbon black that does not permit passage of most solar radiation. Rather, it absorbs solar radiation and this rescues the heating of soil by several degrees. However, it is more stable and lasts considerably longer under field conditions (Streck et al. 1995).

Several studies have evaluated the effect of plastic thickness on solarization efficiency (Saleh et al. 1990, Habeeburrhman and Hosmani 1996, Marenco and Lustosa 2000, Mudalagiriyappa et al. 1999). In Jordan, Saleh et al. (1990) found that black plastic with medium thicknesses (60–80 μm) increased soil temperature than thinner (40, 60 μm) and thicker plastics (100, 120, and 200 μm). In addition, Mudalagiriyappa et al. (1999) showed that solarization with 50 μm transparent polyethylene attained greater pod yield in kharif groundnut in India than 75 μm of the same plastic and than black sheets. Habeeburrhman and Hosmani (1996) concluded that solarization with 50 mum transparent polyethylene resulted in greater grain yield in kharif sorghum and greater reduction in weed biomass than 100 μm transparent and black sheets.

In addition to decreasing the germination percentage the solarization with thin transparent PE sheets significantly reduced the germination rate for seed solarized at both 2.5 and 15 cm depths for 30 and 45 days (Fig. 1 and Table 4). The germination rate for seed solarized at 2.5 cm for 45 days was 12.5, compared to 44.9 for no-solarized seeds. The lower rate of germination indicates that seedling emergence in solarized soil will be delayed behind the emergence of the grown crops. Consequently, the emergent seedling of *P. oleracea*, especially those from surface layer, would face soil dryness and hence greater mortality following their emergence.

The deterioration in seed germination was significantly greater for seeds of surface soil layer (2.5 cm) than the deeper depth (15 cm). This indicates that solarization would help in greater deterioration of the soil seed back in the surface layer than the deeper layer. Consequently, our results recommend to prepare the soil for cultivation prior to solarization process and to avoid any plowing or any kind of soil disturbance that would bring the viable seeds of the deep layer up to the soil surface.

The effectiveness of soil solarization on weed seeds varies with seed type and size, soil depth, plastic type, soil moisture, and duration of solarization (Elmore, 1991, Linke, 1994). The results showed a considerable proportion of *P. oleracea* seeds resisted solarization for 45 days at 15 cm depth. This proportion varied between 26% under thin transparent sheets to 45% under black sheets (Fig. 1). Elmore (1991) indicated that *Portulaca* species were among the most resistant summer weeds to soil solarization. This species is often used as an indicator species such that its control with solarization will often result in control of most weed species (Egley 1983, 1990, Duranti and Cuocolo 1988, Abu-Irmaileh 1991). El-Keblawy et al. (2004) studied the effect of soil solarization on seed viability of three annual weeds under the environment of the UAE and found *P. oleracea* was the most resistant weed. Despite 19 days of solarization were enough to kill all the seeds of *Capsella bursa-pastor* and *Amaranthus graecizans*, the same period increased the germination of *P. oleracea* at 10 cm depth over that of the control (El-Keblawy et al. 2004). These results strongly recommend the use of thin transparent instead of black plastic, which is regularly used in the UAE. The cheaper prices of the thin transparent plastic would encourage farmers to use solarization, which is an environmentally friendly method in disinfestation of soil-borne pests and weeds in a wide scale in the Arabian Gulf countries as they receive intensive solar energy throughout the year.

REFERENCES

Abu-Irmaileh B.E. (1991). Weed control in squash and tomato fields by soil solarization in the Jordan Valley. *Weed Research* 31, 125–133.

Ahmad, Y., Hameed, A. & Aslam, M. (1996). Effect of soil solarization on corn stalk rot. *Plant and soil* 179: 17–24.

Al-Masoum, A.A., Saghir, A. & Hani, S. (1993). Soil solarization for weed management in UAE. *Weed Technology* 7: 507–510.

Al-Masoum, A.A., Ahmed, A., Hashim, K., Jaafer & Al Asool, A. (1998). Effect of two mulch types for solarization on soil temperature. *Agriculture Mechanization in Asia, Africa and Latin America* 29: 73–75.

Arora, A. & Yaduraju, N.T. (1998). High temperature effects on germination and viability of weed seeds in soil. *Journal of Agronomy and Crop Science* 181: 35–43.

Brighton, G.A. (1972). Degradation and disposability of plastic. In: *Plastics and Agriculture*, Vol 1: 203–208. Fifth International Colloquium, Budapest.

Chellemi, D.O., Rhoads, F.M., Olson, S.M., Rich, J.R., Murray, D., Murray, G. & Sylvia, D.M. (1999). An alternative, low-input production system for fresh market tomatoes. *American Journal of Alternative Agriculture* 14:59–68.

Devay, J.E. (1991). Use of soil solarization for control of fungal and bacterial plant pathogens indicating bio-control. *Plant Production and Protection, Paper No. 109, Rome, Italy*, p. 79–84.

Duranti, A. & Cuocolo, L. (1988). Solarization in weed control for onion (*Allium cepa* L.). *Advances in Horticultural Science*, 2, 104–108.

Egley, G.H. (1983). Weed seed seedling reductions by soil solarization with transparent polyethylene sheets. *Weed Science* 31:404–409.

Egley, G.H. (1990). High-temperature effects on germination and survival of weed seeds in soil. *Weed Science*, 38, 42–435.

El-Keblawy, A., Dhaheri, H., Al Mazrooei, S. S., Ghaili, N. A., Al Hedhani, E. M., Al Salmi, A.M., Ali, R.R.M. & Al Ameri, H. S. (2004). Using Soil Solarization in Weed Control under the Environment of the UAE. In: *Proceeding of Fifth Annual Conference for Research Funded by UAE University*, Al-Ain, UAE.

El-Keblawy, A. (2002). Effects of soil solarization on seed germination of fresh and stored seeds of Amaranthus viridis under the environment of the UAE In*: Proceedings I of "Nineteenth Asian-Pacific Weed Science Society Conference"* (Casimero, M.C., Janiya, J. D. &. Pablico, S. Ma, eds.) p. 379–386, Publication of Bureau of Agricultural Research, Manila, Philippines.

Elmore, C.L. (1991). Weed Control by Solarization. *In*: Katan, J., Devay, J.E. (eds.), Soil Solarization, CRC-Boca Ration, Fla, p. 61–72.

Habeeburrhman, P.V., Hosmani, M.M. (1996). Effects of soil solarization in summer on weed growth and yield of succeeding rainy-season sorghum (*Sorghum Bicolor*), *Indian Journal of Agronomy* 41: 54–57.

Holm, L.G., Plucknett, D.L., Pancho, J.V., & Herberger, J.P. (1977). *The World's Worst Weeds: Distribution and Biology,* University Press of Hawaii.

Katan, J. (1981). Solar heating (solarization) of soil for control of soil borne pests. *Annual Review of Phytopathology* 19: 211–236.

Khan, M.A. & Ungar, I.A. (1984). The effect of salinity and temperature on the germination of polymorphic seeds and growth of *Atriplex triangularis* Willd. *American Journal of Botany* 71: 481–489.

Linke, K.H. (1994). Effects of soil solarization on arable weeds under Mediterranean conditions: control, lack of response or stimulation. *Crop Protection*, 13: 115–120.

Marenco, R.A., Lustosa, D.C. (2000). Soil solarization for weed control in carrot . *Pesquisa Agropecuaria Brasileira* 35: 2025–2032

Mauromicale, G., Restuccia, G. & Marchese, M. (2001). Soil solarization, a non-chemical technique for controlling *Orobanche crenata* and improving yield of faba bean. *Agronomie* 21: 757–765.

Mudalagiriyappa, A., Nanjappa, H.V. & Ramachandrappa, B.K. (1999). Effect of soil solarization on weed growth and yield of kharif groundnut (*Arachis hypogaea*). *Indian Journal of Agronomy* 44: 396–399.

Ricci, M.S.F., de Almeida, D.L., Ribeiro, R.D.D., de Aquino, A.M., Pereira, J.C., De-Polli, H., Reis, V.M. & Eklund, C.R. (1999). *Cyperus rotundus* control by solarization. *Biological Agriculture & Horticulture* 17:151–157.

Saleh, H., Abu-Gharbieh, W.I. & Abu-Blan, H. (1990). Effect of solarization using different thicknesses of black plastic tarping on soil-borne pathogens. *Pure and Applied Sciences* 17: 1990. 41–53.

Schreiner, R.P., Ivors, K.L. & Pinkerton, J.N. (2001). Soil solarization reduces arbuscular mycorrhizal fungi as a consequence of weed suppression. *Mycorrhiza*. 11(6):273–277.

Singh, K.P. (1973). Effect of temperature and light on seed germination of two ecotypes of *Portulaca oleracea* L. *New Phyto*logist 72: 289–295.

Stapleton, J.J. & Devay, J.E. (1986). Soil solarization: a non-chemical approach for management of plant pathogens and pests. *Crop Protection* 5: 190–198.

Stapleton, J.J. (2000). Soil solarization in various agricultural production systems. *Crop Protection* 19: 837–841.

Streck, N.A., Schneider, F.M., Buriol, G.A & Heldwein, A.B. (1995). Effect of polyethylene mulches on soil temperature and tomato yield in plastic greenhouse. *Scientia Agricola* 52: 587–593.

Zimmerman, C.A. (1976). Growth characteristics of weediness in *Portulaca oleracea* L. *Ecology,* 57, 964–974.

Arid Land Hydrogeology: In Search of a Solution to a Threatened Resource – Mohamed(ed.)
© 2006 Taylor & Francis Group, London, ISBN 0 415 41127 0

Author index